RECONFIGURABLE SYSTEM DESIGN AND VERIFICATION

RECONFIGURABLE SYSTEM DESIGN AND VERIFICATION

Pao-Ann Hsiung
Marco D. Santambrogio
Chun-Hsian Huang

CRC Press
Taylor & Francis Group
Boca Raton London New York

CRC Press is an imprint of the
Taylor & Francis Group, an **informa** business

CRC Press
Taylor & Francis Group
6000 Broken Sound Parkway NW, Suite 300
Boca Raton, FL 33487-2742

Printed in the United States of America on acid-free paper
10 9 8 7 6 5 4 3 2 1

International Standard Book Number-13: 978-1-4200-6266-3 (Hardcover)

Library of Congress Cataloging-in-Publication Data

Hsiung, Pao-Ann.
Reconfigurable system design and verification / Pao-Ann Hsiung, Marco D. Santambrogio, and Chun-Hsian Huang.
p. cm.
Includes bibliographical references and index.
ISBN 978-1-4200-6266-3 (hardcover : alk. paper)
1. Embedded computer systems. 2. System design. 3. Computer systems--Verification. I. Santambrogio, Marco D. (Marco Domenico) II. Huang, Chun-Hsian. III. Title.

TK7895.E42H79 2009
004.2'1--dc22 2008044105

Visit the Taylor & Francis Web site at
http://www.taylorandfrancis.com

and the CRC Press Web site at
http://www.crcpress.com

To the two greatest ladies in my life:
my wife Nancy and my mother Chin-Shong

– Pao-Ann Hsiung

To my wife Emanuela and to my beloved family

– Marco D. Santambrogio

To my beloved family

– Chun-Hsian Huang

Contents

Preface

As has been widely studied and proved, many emerging applications in communication, computing, and consumer electronics demand that their functionality stay flexible after the system has been manufactured; therefore, the design of embedded systems has changed rapidly during the last decade introducing or defining the reconfigurable computing field as a new direction to enlarge the system solution space. In line with this observation, we now need to focus on how reconfigurable computing can be integrated into our system design using a complete methodology that allows us to easily implement on an FPGA (Field Programmable Gate Array), the best-known and most widely used reconfigurable chip at this moment, a system specification, taking as input its high-level description, such as SystemC or C, and exploiting the capabilities of partial dynamic reconfiguration and hardware/software codesign methodologies.

The demand for reconfigurable computing is growing at an amazing speed. However, we still cannot find many books that teach us all the related techniques for the design and verification of reconfigurable systems. The experiences of experts have been confined to their own application domains. Students and engineers lack the required training for developing such systems and there is no single source of references. A general awareness of the need to evolve our current computer science and electrical engineering curriculum has been growing with the pervasiveness of reconfigurable computing. This book strives to bridge the gap between the need for reconfigurable computing education and the burgeoning development of numerous different techniques in the design and verification of reconfigurable systems in various application domains.

Readership

This book can be used as both a textbook and a reference book. The textbook can be used for courses such as reconfigurable computing, reconfigurable system design, FPGA systems, and hardware-software co-design. The examples, case studies, and exercises provided with each chapter illustrate the theory behind the techniques and thus help students and readers to put the knowledge into practice. The reference book can be used by hardware engineers, software engineers, system engineers, project managers, and verification engineers related to reconfigurable systems. The book can be used as a single source of techniques for the design and verification of any reconfigurable application under design. These techniques otherwise would have been dispersed in the literature making it very difficult to search, compare, and select.

Organization

The book focuses on the individual techniques used in the design and verification of reconfigurable systems and not on individual applications or frameworks. The aim of the book is to make the techniques available to different applications as required by the designer. The book is divided into seven chapters, starting with an introduction to reconfigurable computing and concluding with a detailed design and implementation flow. The seven chapters provide an introduction to reconfigurable systems, FPGA technology with dynamic reconfiguration, reconfigurable hardware design techniques, operating system design for reconfigurable systems, dynamic reconfigurable system design flows, reconfigurable system verification techniques, and reconfigurable system design implementation techniques.

Contents

In Chapter 1, the introduction to reconfigurable computing, we set the background for reconfigurable system design by introducing reconfiguration techniques and architectures, tools and platforms, and design and verification methodologies. Application examples are given to illustrate the techniques and tools.

In Chapter 2, FPGA technology, we introduce the various types of reconfiguration architectures with emphasis on the FPGA, including its configuration bitstream, chip families, configuration conventions, file formats, and different reconfiguration techniques.

In Chapter 3, the reconfigurable hardware design techniques, we first describe the model for reconfigurable hardware and then describe in detail the partitioning and scheduling techniques for reconfigurable hardware.

In Chapter 4, the operating system for reconfigurable systems, we describe the motivation and the requirements for such an OS and then describe a layered architecture for the OS, including the hardware layer, configuration layer, placement layer, scheduling layer, module layer, and application layer. This layering is recommended for any operating system that is targeted for reconfigurable systems.

In Chapter 5 on the reconfigurable system design flows, we first describe the basic and generic flows for system design and then go into details of the design flow, including structure and implementation. The hardware side of the design flow is then discussed.

In Chapter 6 on reconfigurable system verification techniques, we describe system-level verification techniques including formal verification and the language approach then go into the details of hardware-software co-simulation and prototyping. Finally, we discuss frameworks with special focus on Perfecto.

Chapter 7 tries to bridge all the other chapters with a detailed account of the method by which reconfigurable systems can be designed and implemented using Xilinx FPGA chips, tools, platforms, and flows.

Platforms and Tools

The design and prototyping platforms and tools used in the book are mainly from Xilinx. For dynamic partially reconfigurable systems, Xilinx Virtex II Pro FPGA or Virtex 4 FPGA, and embedded PowerPC microprocessors are recommended for implementing the examples given in the book. For other full and statically reconfigurable systems, both Xilinx and Altera FPGA platforms can be used.

Exercises and Projects

Exercises are provided at the end of each chapter for students to familiarize themselves with the techniques described in the chapter. Projects related to the topics discussed in the chapter are also listed at the end of the exercises.

Teaching Supplements

The web page for the book (http://www.cs.ccu.edu.tw/ pahsiung/RSDVbook) contains slides for teaching, model course syllabi, and all source code used in the book. Solutions to the exercises will be available only for instructors through the publisher.

Suggestions

We welcome all kinds of suggestions and comments on the book, including any errors or omissions that you identify. Please e-mail to `rsvd-book@embedded.cs.ccu.edu.tw`.

Acknowledgments by Pao-Ann Hsiung

The book could not have been developed without the efforts of students taking related courses and doing research projects at the National Chung Cheng University on the topic of reconfigurable systems. There are already 13 master theses published and more than 10 PhD dissertations to be published in the coming few years on this topic. All of these graduate students have done remarkable work and graduated or will be graduating from the embedded systems laboratory of National Chung Cheng University, Taiwan. The projects were all sponsored by the National Science Council, Taiwan.

Further, I would like to thank my wife, Nancy, for her kind help and consideration during the long span of time spent on writing this book. I would like to say a big "sorry" to my children, Alice, Phoebe, Roger, and Tiffany, all of whom were neglected by me for long periods of time. This book could not have materialized without their help and kindness.

Last but not the least, I would like to thank my colleagues, Professors Trong-Yen Lee, Mao-Hsu Yen, and Jih-Ming Fu, for their active cooperation in developing related course materials and in the course promotion project related to reconfigurable system prototyping and sponsored by the Ministry of Education, Taiwan.

Acknowledgments by Marco D. Santambrogio

First of all, I would like to thank Emanuela, my wife. She has been so sweet and understanding during these months, without her, I would never have been able to complete this work. My family has helped me during my whole life.

I would like to thank D. Sciuto and all the DRESD guys: they are simply terrific! They helped me believe that something extraordinary was possible, and that's true because of them! I especially want to thank: A. Antola, C. Bolchini, F. Bruschi, F. Cancare', D. Candiloro, R. Cordone, M. Giani, F. Ferrandi, A. Miele, A. Montone, M. Morandi, M. Murgida, S. Ogrenci Memik, M. Novati, A. Panella, V. Rana, F. Redaelli, M. Redaelli, C. Sandionigi, F. Sironi, and M. Triverio. I appreciate all their ideas. The discussions we had together helped with this book. I'm sure that I'm forgetting someone, but that's why I'd like to add this acknowledgment to all of you, who are now reading this page!
Thanks to all of you!

Acknowledgments by Chun-Hsian Huang

First I would like to thank my advisor, Prof. Hsiung. He spent much time supervising me, which helped me acquire related knowledge and write this book. Additionally, I am very thankful to Marco, who is a very experienced researcher. I was very happy to be able to co-write with him and my advisor. I think the writing for this book has been my best experience in my research.

I also wish to thank all of the members of the embedded systems laboratory of National Chung Cheng University, Taiwan. Each of them spent much time doing reconfigurable computing research so that we could achieve the abundant research results. All of them very kindly helped me complete the writing.

Finally, I have to thank my family. When I started my master's and PhD programs, I spent much time on my research and little time at home. I am very thankful for their support. I am extremely hopeful that this book will be very helpful to anyone who is interested in the reconfigurable computing.

Pao-Ann Hsiung
Department of Computer Science and Information Engineering
National Chung Cheng University, Taiwan, ROC.
Marco D. Santambrogio
Dipartimento di Elettronica e Informazione
Politecnico di Milano, Milano, Italy
Chun-Hsian Huang
Department of Computer Science and Information Engineering
National Chung Cheng University, Taiwan, ROC.

Author Biographies

Pao-Ann Hsiung, Ph.D., received his B.S. in Mathematics and his Ph.D. in Electrical Engineering from the National Taiwan University, Taipei, Taiwan, ROC, in 1991 and 1996, respectively. From 1993 to 1996, he was a Teaching Assistant and System Administrator in the Department of Mathematics, National Taiwan University. From 1996 to 2000, he was a post-doctoral researcher at the Institute of Information Science, Academia Sinica, Taipei, Taiwan, ROC. From February 2001 to July 2002, he was an assistant professor and from August 2002 to July 2007 he was an associate professor in the Department of Computer Science and Information Engineering, National Chung Cheng University, Chiayi, Taiwan, ROC. Since August 2007, he has been a full professor.

Dr. Hsiung was the recipient of the 2001 ACM Taipei Chapter Kuo-Ting Li Young Researcher for his significant contributions to design automation of electronic systems. This award is given annually to only one person under the age of 36, conducting research in Taiwan. Dr. Hsiung was also a recipient of the 2004 Young Scholar Research Award given by National Chung Cheng University to five young faculty members per year.

Dr. Hsiung is a senior member of the IEEE, a senior member of the ACM, and a life member of the IICM. He has been included in several professional listings such as Marquis' Who's Who in the World (starting from the 17th Millenium Edition, 2000), Marquis' Who's Who in Asia (starting from the 1st Edition, 2007), Outstanding People of the 20th Century (2nd Edition, 2000), 2000 Outstanding People of the 21st Century (1st edition, 2008), by International Biographical Centre, Cambridge, England, Rifacimento International's Admirable Asian Achievers (2006), Afro/Asian Who's Who, Vol. I (2007), and Asia/Pacific Who's Who, Vol. VII, (2007), Who's Who in Formal Methods, ACM SIGDA's design automation professionals, etc. Dr. Hsiung is an editorial board member of the *International Journal of Embedded Systems* (IJES), Inderscience Publishers, USA; the *International Journal of Multimedia and Ubiquitous Engineering* (IJMUE), Science and Engineering Research Center (SERSC), USA; an associate editor of the *Journal of Software Engineering* (JSE), Academic Journals, Inc., USA; an editorial board member of the *Open Software Engineering Journal* (OSE), Bentham Science Publishers, Ltd., USA; an international editorial board member of the *International Journal of Patterns* (IJOP); and has guest edited two special issues in 2005 and 2006 for IJES. Dr. Hsiung has been on the program committee of more than 50 international conferences. He served as session organizer and chair for PDPTA'99, and as workshop organizer and chair for RTC'99, DSVV'2000, and PDES'2005. He has published more than 150 papers in international journals and conferences. He has taken an active part in paper refereeing for international journals and conferences.

His main research interests include reconfigurable computing and system design, multi-core programming, cognitive radio architecture, System-on-Chip (SoC) design and verification, embedded software synthesis and verification, real-time system design and verification, hardware-software codesign and coverification, and component-based object-oriented application frameworks for real-time embedded systems.

Marco D. Santambrogio, Ph.D., received his laurea (M.Sc. equivalent) degree in Computer Engineering from the Politecnico di Milano in 2004, his second M.Sc. in Computer Science from the University of Illinois at Chicago (UIC) in 2005, and his Ph.D. in Computer Engineering from the Politecnico di Milano in 2008. He has held visiting positions at the Department of Electrical Engineering and Computer Science of Northwestern University (2006 and 2007) and Heinz Nixdorf Institut (2006). He is affiliated with the Micro Architectures Laboratory (MicroLAB) at the Politecnico di Milano, where he founded the Dynamic Reconfigurability in Embedded System Design (DRESD) project in 2004. He conducts research and teaches in the areas of reconfigurable computing, hardware/software codesign, embedded systems, and high performance processors and systems. He has been involved in teaching activities in Dipartimento di Elettronica e Informazione, Politecnico di Milano, since 2004, in Universita' degli Studi di Milano, since 2005, and at ALaRI, Advanced Learning and Research Institute, University of Lugano, since 2005.

Dr. Santambrogio is a member of the IEEE, the IEEE Computer Society (CS) and the IEEE Circuits and Systems Society (CAS). He is guest editing a special issue for the *EURASIP Journal of Embedded Systems*. He has been a reviewer for *IEEE Transactions on Very Large Scale Integration Systems*, *ACM Transaction on Reconfigurable Technology and Systems*, *Journal of Systems Architecture* and different international conferences. He has been on the program committee of several international conferences. He served as session organizer and chair for IEEE International Conference on Field Programmable Logic and Applications (FPL), IEEE Reconfigurable Architectures Workshop (RAW), IEEE Computer Society Annual Symposium on VLSI (ISVLSI), International Conference on ReConFigurable Computing and FPGAs (ReConFig), IEEE Southern Conference on Programmable Logic (SPL), IEEE Field Programmable Technology (FPT), and Engineering of Reconfigurable Systems and Algorithms (ERSA). Since 2001, he has been involved in several research projects in collaboration with industries such as ATMEL, Siemens Mobile, and Nokia Siemens Network.

Chun-Hsian Huang received his B.S. degree in Information and Computer Education from National TaiTung University, TaiTung, Taiwan, ROC, in 2004. He is currently working toward his Ph.D. in the Department of Computer Science and Information Engineering at National Chung Cheng University, Chiayi, Taiwan, ROC. He is a teaching and research assistant in the Department of Computer Science and Information Engineering at National Chung Cheng University. His research interests include dynamically partially reconfigurable systems, UML-based hardware/software co-design methodology, hardware/software co-verification, and formal verification.

1

Introduction to Reconfigurable Computing

Progress in technology induces paradigm shifts in computing. The invention of the programmable microprocessor in 1974 resulted in the evolution from pure hardware-based computing to software-based computing. With the advent of programmable hardware in the early 1980s, we have configware-based computing. Finally, in the 21st century, due to Moore's Law, the number of gates that can be squeezed into a tiny chip area has increased to such an extent that a full hardware-software system can be implemented on a single chip. With the advent of System-on-Chip (SoC), the design of electronic consumer products initiated an era of digital convergence, leading to the integration of numerous applications onto a single device. The mobile phone is a typical example, integrating GPRS, GSM, GPS, MP3, MPEG4, CCD Digital Still Camera, FM Radio, Wi-Fi, Bluetooth – you name it and you have it on your mobile phone!

This crowded integration has also led to high power consumption, high design cost, large form factor, low mean-time between failure, low reliability, increased security risks, and a lot of side effects and conflicts. Many of these issues can be resolved if reconfigurable logic is introduced into the system. In simple terms, *reconfigurable logic* is a special kind of hardware circuit that can be reconfigured, after fabrication, into whatever logic the user desires. The reconfiguration process is often simply *programming* some kind of configuration memory. There are already mobile phones on the market that have reconfigurable logic embedded into the system. However, consumer products have not yet all followed this trend, which has, however, swept the scientific large-scale computing world.

Besides alleviating the above-mentioned problems, reconfigurable computing is breaking down the barrier between hardware and software design technologies. The segregation between the two has become more and more fuzzy because both are programmable now. Reconfigurable computing can also be viewed as a trade-off between general-purpose computing and application specific design. Given the architecture and design flexibility, reconfigurable computing has catalyzed the progress in hardware-software codesign technology and in a vast number of application areas such as scientific computing, biological computing, artificial intelligence, signal processing, security computing, and control-oriented design, to name a few.

In the rest of this chapter, we first briefly introduce why and what is reconfigurable computing. We then go on to describe the resulting enhancements of hardware-software codesign methods and the techniques, tools, platforms, and design and verification methodologies of reconfigurable computing. Furthermore, we will introduce and compare some reconfigurable computing architectures. Finally, the future trends and conclusions will also be presented. This chapter is aimed at a wide audience, including both technical persons and those not particularly well grounded in computer architecture.

1.1 Why Reconfigurable Computing?

Traditionally, computing was classified into general-purpose computing performed by a *General-Purpose Processor* (GPP) and application-specific computing performed by an *Application-Specific Integrated Circuit* (ASIC).

General-purpose computing was first accomplished by the electronic general-purpose computer, called the *Electronic Numerical Integrator and Calculator* (ENIAC), built by J. Presper Eckert and John Mauchly. But it is well known as the von Neumann computer because ENIAC was improved by John von Neumann [84]. A general-purpose computer is a single common piece of silicon, called a microprocessor or GPP, that can be programmed to solve any computing task. This means many applications could share commodity economics for the production of a single integrated circuit (IC). This computing architecture has the flexibility and superiority that the original builders of the IC never conceived [157].

An ASIC is an IC specifically designed to provide unique functions. ASIC chips can replace general-purpose commercial logic chips, and integrate several functions or logic control blocks into one single chip, lowering manufacturing cost and simplifying circuit board design. Although the ASIC has the advantages of high performance and low power, its fixed resource and algorithm architecture result in drawbacks such as high fabrication cost and poor flexibility.

As a trade-off between the two extreme characteristics of GPP and ASIC, reconfigurable computing has combined the advantages of both. A comparison of the characteristics of the three different architectures is given in Table 1.1 [189, 187].

From Table 1.1, we observe that reconfigurable computing has the combined advantages of configurable computing resources, called *configware* [106], as well as configurable algorithms, called *flowware* [107, 81]. Further, the performance of reconfigurable systems is better than general-purpose systems and the cost is smaller than that of ASICs. Only recently the power consumption of reconfigurable systems has been improved such that it is now either comparable with ASICs or even smaller. The main advantage of the reconfigurable system is its high flexibility, while its main disadvantage is the lack

Table 1.1: Comparison of Representative Computing Architecture

Architecture	General Purpose	ASIC	Reconfigurable
Resources	Fixed	Fixed	Configware
Algorithms	Software	Fixed	Flowware
Performance	Low	High	Medium
Cost	Low	High	Medium
Power	Medium	Low	Medium
Flexibility	High	Low	High
Computing Model	Mature	Mature	Immature
NRE Cost	Low	High	Medium

of a standard computing model. The design effort in terms of non-recurring engineering (NRE) cost is between that of general-purpose processors and ASICs.

We can thus conclude that reconfigurable computing is a trade-off between general-purpose computing and application-specific computing because it tries to achieve a balance among performance, cost, power, flexibility, and design effort. Reconfigurable computing has enhanced the performance of applications in a large variety of domains, including embedded systems, SoCs, digital signal processing (DSP), image processing, network security, bioinformatics, supercomputing, Boolean SATisfiability (SAT), spacecrafts, and military applications. We can say that reconfigurable computing will widely, pervasively, and gradually impact human lives. Hence, it is time that we focus on how reconfigurable computing and reconfigurable system design techniques are to be utilized for building applications.

1.2 What is Reconfigurable Computing?

In 1960, Estrin [62] was the first to propose the term "reconfigurable computing"; however, it gained industrial acceptance only after the advent of *Field Programmable Gate Arrays* (FPGAs) in the 1980s. Further, it was only recently that the nanometer deep submicron fabrication technology allowed a full SoC to be implemented using reconfigurable technology.

As illustrated in Figure 1.1, a reconfigurable computing architecture is composed of a general-purpose processor and some reconfigurable hardware logic. The reconfigurable hardware logic executes application-specific tasks that are computation intensive, such as the encryption task (C_1) and the image processing task (C_5). The GPP is used to control the behavior of the tasks running in the reconfigurable hardware logic and is responsible for functions such as external I/O communications. As shown in Figure 1.1, when a re-

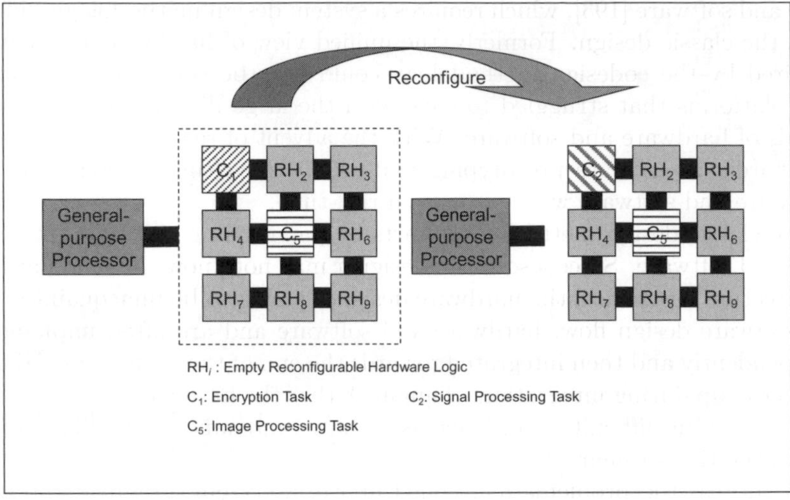

Figure 1.1: Reconfigurable computing.

configurable hardware task, such as the encryption task C_1, has finished its computation the processor reconfigures the hardware logic to execute another task, such as the signal processing task C_5. During this reconfiguration process, the image processing task C_5 continues to execute, without interruption. The reconfigurable computing architecture can also be concisely described as *Hardware-On-Demand* [164], general purpose custom hardware [74], or a hybrid approach between ASIC and GPP [170].

From the above illustration, we can also define reconfigurable computing as a discipline in which the functions of a system or an application can be altered by configuring a fixed set of logic resources through memory settings. Examples of application functions include various kinds of transforms, filters, codecs, and protocols. Examples of the fixed set of logic resources include logic blocks, I/O blocks, routing blocks, memory blocks, and application-specific blocks. Memory settings can be achieved by the programming of configuration bits that control the functions of the logic resources.

1.3 From Codesign to Reconfiguration

To understand the development and benefits of reconfigurable computing, an important perspective is to view it from the transition of the hardware-software (HW-SW) codesign technology to the reconfigurable computing technology. Codesign is an emerging topic that highlights a unified view of hardware and software [193], which requires a system design methodology different from the classic design. Formerly, the unified view of hardware and software required by the codesign methodology could only be provided by cosimulation platforms that struggled to cope with the large difference in simulation speeds of hardware and software. With the advent of reconfigurable computing, hardware-software prototyping platforms solved this difference as both hardware and software were running in real-time.

The classic design methodology starts by partitioning a system into hardware and software. Since a software designer may not know the final hardware architecture design and the hardware designer may also be unacquainted with the software design flow, hardware and software and are often implemented independently and then integrated towards the end of the design flow. If problems crop up during integration, changing either the hardware or the software could be quite difficult. This increases the maintenance difficulty and also delays the time-to-market.

To address the problems mentioned above, a system design methodology called HW-SW codesign was proposed, which emphasizes the consistency and the integration between hardware and software. It is based on a system-level view and thus eases both software verification and hardware error detection. The HW-SW codesign methodology reduces the cost of design and also shortens the time-to-market.

Nevertheless, the high cost in hardware design is a major issue of the HW-SW codesign flow because hardware must go through a time-consuming flow including design, debug, manufacturing, and test. The inconvenient hardware manufacturing forces designers to search for alternate ways. One way is to use modeling languages such as SystemC [23] to simulate hardware and software. Another method is to use concurrent process models to simulate hardware and software tasks. However, simulation speed is a major bottleneck [164]. To overcome the drawback, prototyping using reconfigurable architectures has become the most appropriate choice for HW-SW codesign. In contrast to ASIC design, reconfigurable hardware can be much more easily used to design hardware prototypes that can be integrated with software to obtain performance and functional analysis results much more efficiently and accurately. We can thus say that reconfigurable computing has accelerated and enhanced the HW-SW codesign flow. Figure 1.2 compares classic design, hardware-software codesign, and reconfigurable design. There is an obvious schedule compaction in the system design flow when transiting from classic design to co-design and then from co-design to reconfigurable design. The first transition mainly results in hardware and software being designed concurrently and thus reducing the design time. The second transition results in a compaction of the system design, which previously required lengthy hardware-software cosimulation time.

In the rest of this chapter, we will first introduce the technology and the devices used for reconfiguration, including the famous *Field-Programmable Gate Arrays* (FPGA) [73] in Section 1.4. We will then introduce some represen-

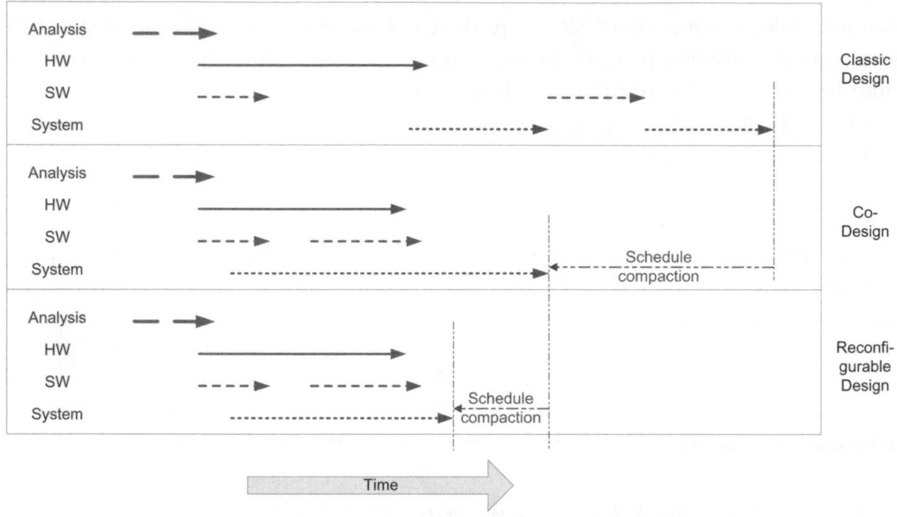

Figure 1.2: Classic design vs. co-design vs. reconfigurable design.

tative reconfiguration tools and platforms used in academia and industry in Section 1.5. Contemporary design and verification methodologies will also be introduced in Section 1.6. Finally, we conclude this chapter by giving some application examples for reconfigurable computing in Section 1.7.

1.4 Reconfiguration Technology

Since the mid-1980s, reconfigurable computing has become a popular field due to the FPGA technology progress. An FPGA is a semiconductor device containing programmable logic components and programmable interconnects [45] but no instruction fetch at run time, that is, FPGAs do not have a program counter [81]. In most FPGAs, the logic components can be programmed to duplicate the functionality of basic logic gates or functional intellectual properties (IPs). FPGAs also include memory elements composed of simple flip-flops or more complete blocks of memories [19]. Hence, FPGA has made possible the dynamic execution and configuration of both hardware and software on a single chip. A detailed description of FPGA can be found in Chapter 2.

Besides FPGA, the *reconfigurable Data-Path Array* (rDPA) proposed by Rainer Kress in 1993 at TU Kaiserslautern [81] is another reconfiguration technology. In contrast to the fine-grained bit-wise configuration in FPGA, rDPAs have reconfigurable data-path units (rDPUs) that are multiple bit wide configurable, for instance, 32-bit path width is typically found. An rDPA is structurally programmed from configware sources, which are compiled into pipe networks to be mapped onto the rDPA.

The main part of a reconfigurable system is the configware such as FPGA or rDPA. Besides configware, the software is another essential part that can control and thus incorporate the configware into a reconfigurable system. Configware can provide resources for high performance computation, but complex control must be implemented in software. Reconfiguring hardware implies software must also be appropriately reconfigured, thus we need reconfigurable software design, too [45, 197].

1.5 Reconfiguration Tools and Platforms

To construct a reconfigurable computing system, designers need computer-aided design (CAD) tools for system design and implementation, such as a design analysis tool for architecture design, a synthesis tool for hardware construction, a simulator for hardware behavior simulation, and a placement and routing tool for circuit layout. We may build these tools ourselves or we can also use commercial tools and platforms for reconfigurable system design, such as the Embedded Development Kit (EDK) [99] from Xilinx, which is a common development tool. The EDK integrates both the software and the hardware components of a design to develop complete systems [56]. In fact, EDK provides developers with a rich set of design tools, such as Xilinx Platform Studio (XPS), gcc, and Xilinx Synthesizer (XST). It also provides a wide selection of standard peripherals required to build systems with embedded processors, like MicroBlaze or IBM PowerPC [103]. Besides Xilinx EDK, we list commonly used commercial FPGA and *Electronic Design Automation* (EDA) tools in Table 1.2.

After designers build a reconfigurable system, a platform for operating and testing is needed. We can use the platforms developed in the industry or in academia, such as the Caronte Architecture [56] and the Perfecto framework [89, 90] for fine-grained systems and the Kress-Kung Machine [81] for coarse-grained systems. The Caronte Architecture is entirely implemented in the FPGA device and consists of several elements including a processor, some memories, a set of reconfigurable devices, and a configuration controller. It implements a module-based system approach based on an EDK system description and provides a low cost approach to the dynamic reconfiguration problem. Perfecto is a SystemC-based design space exploration framework that can be used for evaluating different partitioning, scheduling, and placement alternatives for a reconfigurable system. Perfecto can also be used to design the corresponding algorithms in an operating system for reconfigurable

Table 1.2: Commercial Reconfiguration Tools

Functionality	Tool Name	FPGA/EDA Company
Design Analysis	PlanAhead	Xilinx
FPGA Suite Tools	ISE Foundation	Xilinx
	Quartus	Altera
	FPGA Advantage	Mentor Graphics
FPGA Synthesizer	Synplify Pro	Synplicity
	FPGA Compiler	Synopsys
	Leonardo Spectrum	Mentor Graphics
	Precision Synthesis	Mentor Graphics
Simulator	ModelSim	Mentor Graphics
	NC SIM	Cadence
	Scirocco Simulator	Cadence
	Spexsim	Verisity
	VCS	Synopsys
	Verilog-XL	Cadence

systems. Further information on an operating system for reconfigurable systems and on Perfecto can be found in Chapters 4 and 6, respectively. The Kress-Kung is a data-stream-based machine, consisting of rDPAs and has no *Central Processing Unit* (CPU) or *Program Counter* (PC).

1.6 Design and Verification Methodologies

To design reconfigurable computing systems, we need some appreciation of the different costs and opportunities inherent in reconfigurable architectures. Currently, most systems are designed based on our past experiences. We can use the design patterns identified and cataloged by DeHone et al. [52]. Each pattern description has a name, intent, motivation, applicability, participants, consequences, implementation, known uses, and related patterns. Design patterns are also cataloged into several classification types, such as patterns for area-time trade-offs and patterns for expressing parallelism. This classification is a good start for constructing reconfigurable systems. In the following, we present a typical design methodology and a typical verification methodology for illustration purposes.

Tseng and Hsiung [190] proposed a *Unified Modeling Language* (UML)-based design flow for *Dynamically Reconfigurable Computing Systems* (DRCS). This design flow is targeted at the execution speedup of functional algorithms in DRCS and at the reduction of the complexity and time-consuming efforts in designing DRCS. The most notable feature of the design flow is a HW-

SW partitioning methodology based on the UML 2.0 sequence diagram, called *Dynamic Bitstream Partitioning on Sequence Diagram* (DBPSD). In DBPSD, partitioning guidelines are included to help designers make prudent partitioning decisions at the class method granularity. The enhanced sequence diagram in UML 2.0 is capable of modeling complex control flows, thus the partitioning can be done efficiently on the sequence diagrams.

After design and implementation, we need to verify that the system design is correct and complete. Correctness means that the design implements its specification accurately. Completeness means that our specification describes appropriate output responses to all relevant input sequences [193]. The verification of reconfigurable systems is a more formidable task than the verification of conventional non-reconfigurable systems because of the added dimensionality of dynamic and partial reconfigurations. Usually, the reconfigurations are dynamically controlled by a configuration controller that is driven by an operating system. A typical verification methodology is to evaluate how the partitioning, scheduling, and placement algorithms implemented in an operating system work for a reconfigurable system. The Perfecto framework [89, 90] described in Section 1.5 is a representative tool for such verifications. Perfecto is an easy-to-use system-level framework that is able to perform rapid explorations of different reconfiguration alternatives based on the user-specified combination of partitioning, scheduling, and placement algorithms. In Perfecto, a system designer can simulate and evaluate the performance and the resource usages of different design alternatives, and detect performance bottlenecks, functional errors, architecture defects, and other system faults at a very early design phase. Problems with scheduling and placement strategies can also be detected by Perfecto. More details on Perfecto can be found in Chapter 6.

The design of reconfigurable hardware requires specific knowledge on FPGA and reconfigurable design tools. Details on reconfigurable hardware design can be found in Chapter 3. Further, the operating system and software must also be specifically designed and implemented for reconfigurable systems. Details on the operating system for reconfigurable systems can be found in Chapter 4. More details on design flows can bc found in Chapter 5. Verification of reconfigurable systems can be found in Chapter 6. Implementation details for hardware and software, including operating systems, can be found in Chapter 7.

1.7 Application Examples

The number of application fields in which reconfigurable computing has been applied is too great to list. Instead, we focus on a few of the most typical and widely applied application domains. In this section, we will discuss how reconfigurable computing has or is being applied, to embedded systems, network security applications, multimedia applications, scientific computing, and SATisfiability solvers. More detailed applications can be found in [73].

1.7.1 Embedded Systems

Transportation vehicles have several embedded systems that are connected by a *Controller-Area Network* (CAN) bus. Besides the embedded safety systems, modern car users require *Global Positioning System* (GPS)-based navigation systems, short-distance radio systems, telecommunication systems, Bluetooth-connected multimedia systems, mobile entertainment systems, and so on and so forth. The number of application requirements is growing at such a rapid rate that their integration incurs an overhead not only on the accumulative weight of the car, but also on the power consumption. Sky-rocketing rates of petroleum gas and other fuels are making the need for low power consumption in the embedded systems of a car all the more necessary.

Reconfiguration technology is already being applied to the *Engine Control Units* (ECUs) of a car [8], where we suppose an ECU is responsible for one embedded system application. Due to reconfigurable computing, not all ECUs need to exist at the same time in the car. Only those ECUs that are responsible for the current application requirements need to be existing and connected to the CAN bus. In this way, power consumption is reduced and the weight of the car is also reduced. These reductions, in turn, save fuel and thus indirectly also help protect the environment from pollution. Reconfiguration technology can thus have far-reaching effects on human and the related industrial technology.

1.7.2 Network Security Applications

Network connectedness among systems of varying application domains has been rapidly growing such that security has become a major concern across the domains. Globalization of the industry and human resources has necessitated the need for more secure exchange of data. However, there are multiple standards for security that use different cryptographic algorithms. For example, besides the currently popular and well-known *Advanced Encryption Standard* (AES), there are still large numbers of systems that use *Data Encryption Standard* (DES) and triple DES. For authentication, besides the popular and widely used *Cyclic Redundancy Code* (CRC), there are also large numbers of data access schemes that employ the *Message Digest* (MD5) algorithm and *Secure Hash Algorithm* (SHA). The large variation in security standards requires security systems to adapt to the different encryption and decryption algorithm requirements dynamically. The dynamically changing requirements, in turn, make security systems difficult and cumbersome to design.

Reconfigurable computing is one way to cope with the dynamic requirements of security systems because we need not connect all hardware and software implementations of all encryption and decryption algorithms to the system bus at the same time [93]. We may just need to allow enough reconfigurable module slots in which the simultaneous executions of cryptographic IPs can satisfy the dynamic requirements. For example, if only two cryptography algorithms are required simultaneously, then only two big enough module slots are necessary.

1.7.3 Multimedia Applications

Multimedia applications are characterized by large amounts of data processing and soft real-time requirements. Further, multimedia standards normally employ very similar kernel technologies, for example, *Discrete Cosine Transform* (DCT), quantization, Huffman encoding, and Variable Length Decoding (VLD) appear in most multimedia standards such as MPEG-4 and H.264. Another typical feature of most multimedia systems is the large variation in processing time for different data streams. For example, an action movie might need more processing time than an art movie because the action movie may have large differences between consecutive frames, while the difference would be quite small in the art movie.

Reconfigurable computing can help not only in accelerating the intensive data processing, but also in reusing the same IPs across different applications [169]. Further, reconfigurable computing can also adapt the performance of a system according to the dynamic requirements of the data streams. For example, for an art movie the DCT module could be a slower 1-dimensional hardware implementation, whereas for an action movie, the DCT module would be a faster 2-dimensional hardware implementation. This dynamic change in DCT modules can be made feasible by reconfigurable computing technologies.

1.7.4 Scientific Computing

Scientific computing is mainly characterized by heavy computation tasks that require efficient number-crunching systems and by rapid changes in the input data to be processed that require real-time adaptation and processing. Weather forecasts, biological computing, and earthquake analysis are some examples of scientific computing that requires high performance and high reliability. Reconfigurable computing provides an economical solution to such scientific computing because of the smaller size and scale of system implementations as compared to a full-fledged solution. A typical example for bioinformatics is the RDisk architecture [122], which is composed of 48 connected nodes containing a hard disk and an FPGA in each node. Filters can be reconfigured for different search strategies.

1.7.5 Reconfigurable SAT Solvers

The SATisfiability (SAT) problem, the first well-known NP-Complete problem [46], has an extremely wide range of practical applications in a variety of engineering areas, including electronic circuits testing, pattern recognition, logic synthesis, etc. [75]. Given a set of clauses over Boolean variables, SAT asks if there is a valuation of the variables such that the set of clauses is satisfied. Since it is an NP-complete problem, lots of heuristics were proposed to solve it. Among the proposed methods, Davis and Putnam's method [51] and its improved DPLL version [50] are the most widely known. There are also numerous pure software, pure hardware, and hardware-software SAT solvers for solving the SAT problems. An excellent survey of reconfigurable hardware SAT solvers can be found in [174].

Summary

This chapter was basically a whirlwind tour of the book, introducing several new concepts that will be covered in more detail in later chapters. References to all the other chapters can be found in this chapter. We discussed why we need reconfigurable computing and what exactly is reconfigurable computing. We also suggested the perspective of viewing the progress in reconfigurable computing through its transition from hardware-software codesign. Later, we introduced the technology, the tools, the platforms, and the methodologies for designing and verifying reconfigurable systems. We concluded this chapter using some application examples.

Exercises and Problems

1. List as many application fields as you can in which reconfigurable computing is applied.

2. Why did the FPGA devices, invented in the mid-1980s, take more than two decades to be used in reconfigurable systems?

3. Differentiate between FPGA and rDPA.

4. Define the term *partial* reconfiguration in your own words.

5. What are the main differences between Xilinx and Altera FPGAs?

6. Is there any difference between the tools and platforms used for the design of reconfigurable systems compared to those for conventional systems?

7. **Project Theme**: Try to accelerate the performance of a conventional software program using a reconfigurable system. For example, you could implement the most time-consuming loop iterations in hardware for an image encoding program. However, note that you need to send and receive the data for the loop execution if it is implemented in the hardware (FPGA).

2

FPGA Technology and Dynamic Reconfiguration

This chapter introduces a family of silicon devices, FPGAs, exploring their features and applications, while giving some basic hints about their architecture. We will discuss the low-level configuration details of these devices, so part of this chapter will be devoted to explore the file type that contains the configuration information to initialize the state of these reconfigurable chips, the bitstream file. For describing the bitstream composition, the details have been aggregated from the three different families of FPGAs examined in this book, namely the Spartan-3, Virtex-II Pro, and Virtex 4 families manufactured by Xilinx Inc.

This chapter will then illustrate a particular technique that has been viable for most recent FPGA devices, Partial Dynamic Reconfiguration. To fully understand what this technique is, the concepts of reconfigurable computing, static and dynamic reconfiguration, and the taxonomy of dynamic reconfiguration itself must be analyzed. In this way partial dynamic reconfiguration can be correctly placed in the set of system development techniques that it is possible to implement on a modern FPGA chip.

Finally, reconfigurable systems, while providing new interesting features in the field of embedded system design, also introduce new problems in their implementation and management. This is particularly true for systems that implement partial self-reconfiguration [24, 110, 184, 86, 202, 165]. In *partial* reconfiguration, only *portions* of the reconfigurable device are involved in the configuration change. *Dynamic* reconfiguration allows the device portions that are not directly involved in the reconfiguration to run without interruption through the reconfiguration process. A commonly adopted approach is the definition of predetermined area portions on the device, *reconfigurable slots*, in which components implementing different tasks from the specification, or *modules*, can be configured. The final part of this chapter presents all the different characteristics of a reconfigurable system, starting from the differences between complete and partial reconfiguration to the definition and the description of partial self-reconfigurable architectures.

2.1 FPGA Overview

FPGAs (*Field Programmable Gate Array*) are a particular family of integrated circuits intended for custom hardware implementation, with the key property of being capable of reconfiguration for an infinite number of times. Currently FPGAS are the state of the art of Programmable Logic Devices (PLD), hence this work has been focused on these particular chips. Reconfiguring an FPGA means changing its functionality to support a new application, and it is equal to having some new piece of hardware, mapped on the FPGA chip, having to implement a new functionality. In other words FPGAs make it possible to have custom-designed high-density hardware in an electronic circuit, with the added bonus of having the possibility of changing it whenever there is the need, even while the whole application is still running. An HDL language such as VHDL or Verilog is used to describe the functionality to be implemented on the device, then the design software of the device manufacturer translates the description of the hardware into a configuration file for the device that can be downloaded on it. These software tools are analogous to those employed in ASIC chip design, in the sense that they convert a hardware specification into an actual netlist that can be synthesized, placed, and routed on an actual piece of hardware such as an ASIC or an FPGA.

The flexibility of having custom, changeable hardware in an application is the factor that has determined the popularity of FPGA devices in a broad range of fields. If the FPGA is paired with a general purpose processor, for example, the most demanding sections of the software can be translated into hardware cores that accelerate program execution, yielding notable speedups in the overall execution, especially when software sections executed serially on the processor can be translated into hardware that can exploit the parallelism of the algorithm. This is a viable technique in high performance computing, where software kernels such as FFT or image processing algorithms are implemented on an FPGA, leaving the rest of the program to execute on a general purpose processor. In [111] the authors show the results of executing the Smith-Waterman algorithm, a sequence alignment algorithm in bioinformatics, on a hybrid platform composed of a general purpose processor and an FPGA. The original program is profiled to locate the most computationally intensive functions, then these functions are translated into hardware using an HDL language and finally communication issues between the processor and the hardware on FPGA are taken into account, trying to eliminate bottlenecks. The result is a significant speedup in the execution time of the algorithm. This methodology is broadly used whenever there is the need for accelerating software running over large sets of data, and is commonly referred to as *hardware-software codesign.*

FPGAs can also be used without a paired microprocessor, by implementing all of the application functionalities in hardware. In this case the hardware implemented on the FPGA covers all the data path from the inputs to the outputs of the application. The advantage in this is that the hardware is easily replaceable by downloading an appropriate configuration file onto the chip, rather than having the circuit physically replaced, and this is a key factor in applications such as network appliances, where a node of the network can be easily reconfigured off-site for network upgrades or maintenance. This is one of the reasons why FPGAs can be chosen over ASICs (*Application Specific Integrated Circuits*) in designing custom hardware.

The number of logic gates on the most recent FPGA models has also opened the way for the implementation of complete systems on chip (SoC) on these devices, given the fact that many FPGAs now have one or more processors directly embedded into the silicon wafer, and that even if those processors are not available, soft core processors can be used. Soft core processors are complete processor IP-Cores described in an HDL and synthesizable on FPGAs. Xilinx provides the *Microblaze* and the *Picoblaze* soft core processors to be used in architectures for their FPGAs.

2.1.1 FPGA Architecture

This section will briefly discuss at the distinctive features of FPGAs, in particular to clarify how they can implement custom hardware via their reconfiguration. The architecture of a generic FPGA is illustrated in Figure 2.1. The FPGAs used throughout this work are all manufactured by Xilinx, so the following architectural details apply to the chips produced by this company. However, the general principles according to which FPGAs are made are the same even across different chip brands. The three main building blocks of an FPGA are logic blocks (CLB), IO blocks (IOB), and communication resources.

Configurable logic blocks (CLBs) are the main components of an FPGA. They can have one or more function generators realized with look-up tables (LUT) that can implement an arbitrary logic function according to their configuration. In these components, the result of the function is stored for every possible combination of the inputs, such that a 4-bit LUT will require 16 memory cells to store the function, no matter its complexity. Around the LUT there is the interconnect logic that routes signals to and from the LUT, implemented using standard logic gates, multiplexers, and latches. Therefore, during the configuration process of an FPGA, the memory inside the look-up tables is written to implement a required function, and the logic around it is configured to route the signals correctly in order to build a more complex system around this basic building block. In Xilinx FPGAs a single CLB contains a set of four slices that in turn contain two look-up tables and the necessary interconnect hardware.

The **input output blocks** (IOBs) have the function of interconnecting the

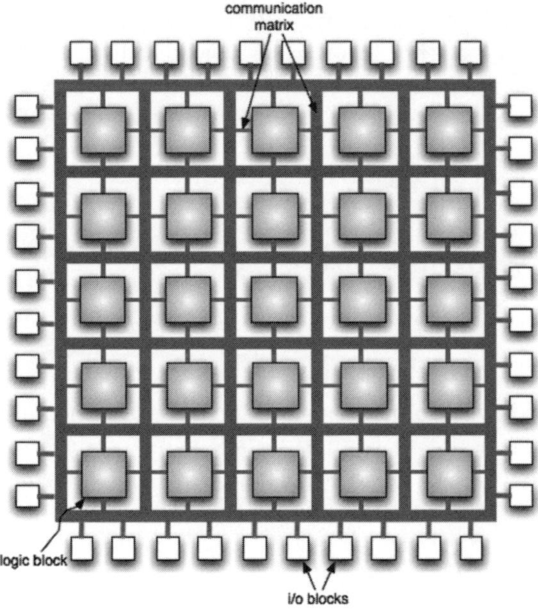

Figure 2.1: Generic architecture of an FPGA.

signals of the internal logic to an output pin of the FPGA package. There is one and only one IOB for every I/O pin of the chip package. The IOBs have their own configuration memory, storing the voltage standards to which the pin must comply and configuring the direction of the communication on it, making it possible to establish mono-directional links in either way or also bidirectional ones.

Finally, the **interconnection resources** within an FPGA allow the arbitrary connection of CLBs and IOBs. The main modes of interconnections are direct and segmented.

- Direct interconnection, see Figure 2.2(a), is made of groups of connections that cross the device in all its dimensions. Logic blocks put data on the most suitable channel according to data destination. This implementation usually includes some additional short-range connections that link to nearby blocks.

- Segmented interconnection is based on lines that can be interconnected using programmable switch matrices. Also in this kind of interconnection there are lines that cross the entire device, in order to maximize the speed of communication and limit signal skew. An illustration can be seen in Figure 2.2(b).

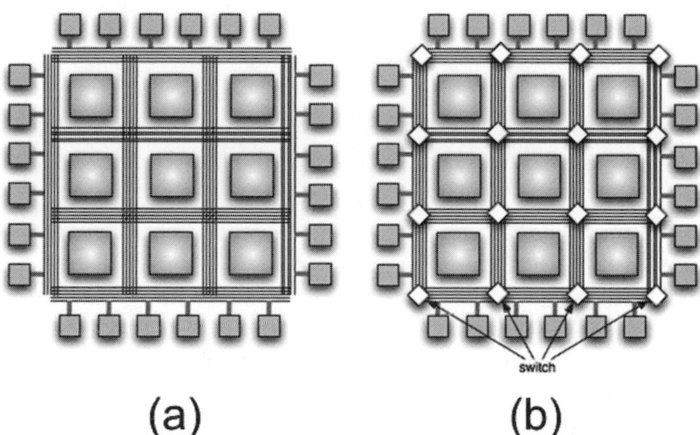

Figure 2.2: FPGA using the direct interconnection model (a) and the segmented interconnection one (b).

The main advantage of direct interconnection is that parasite resistance and capacitance are almost constant, resulting in an improved predictability of the propagation times of the signals. Segmented interconnections offer a reduced power dissipation because resistance and capacity of the interconnection lines are only those of the interconnection length between the blocks.

The three building blocks presented so far interconnect together in the device to create a communication infrastructure composed of the communication lines and IOBs around a bi-dimensional array of CLBs, which covers most of the available die area and represents the true FPGA building block. The memory cells attached to every configurable resource control its key features, in such a way that the IO voltage standards of a IOB are controlled by particular values in its corresponding memory cell, the interconnections among the communication infrastructure are controlled by setting appropriate bits in the configuration memory, and the equations in the LUTS are controlled in the same way. All of this configuration memory is made of SRAM memory elements, and it is therefore volatile: when the device is power-cycled all of its configuration is lost and it must be started afresh with a new configuration. Usually an external machine downloads the configuration on the FPGA via one of its configuration interfaces and sends a start command to signal that the configuration has taken place. Some boards also offer a ROM where configuration is stored, so that it can be subsequently downloaded on the FPGA on power up. The file that stores the information that is copied over the configuration SRAM memory of the FPGA is called *bitstream* and can be either full or partial according to the extent of configuration memory addressed in it.

2.1.1.1 Additional Resources

Most FPGAs are not composed only of the three components described in the previous paragraphs, but they have additional resources directly embedded on the die. Such resources are RAM cells (called *block ram* in Xilinx FPGAs) processors, DSPs, multipliers, and so on. In this way system designers can take advantage of hardware already present on the chip and integrate it in their designs without the need of having to implement all of the desired functionalities on the configurable resources but instead exploiting the speed and the functions of pre-made embedded hardware cores. As an example, the resource array of the Xilinx Virtex II FPGA XC2VP20 [216] is heterogeneous because it is composed of a CLB array interleaved with hard-cores such as an embedded PowerPC processor, 88 blocks of RAM each capable of 18Kb of storage, 8 multi-gigabit transceivers, and so on. A design can therefore be composed of two kinds of hardware: hard-cores and soft-cores. The hard-cores on an FPGA thus enrich its functionalities and improve the overall speed of architectures that make use of them.

2.2 The Configuration Bitstream

A bitstream file* is a binary file in which configuration information for a particular Xilinx device is stored. That is where all the data to be copied on to the configuration SRAM cells are stored, along with the proper commands for controlling the chip functionalities.

Bitstreams can be either partial or full. A full bitstream configures the whole configuration memory of the device, and it is used for static design or at the beginning of the execution of a dynamic reconfiguration system, to define the initial state of the SRAM cells. Partial bitstreams configure only a portion of the device and are one of the end products of any partial reconfiguration flow. FPGAs provide different means for configuration, under the form of different interfaces to the configuration logic on the chip. There are several modes and interfaces to configure a specific FPGA family, among them the JTAG download cable (which is the method used in this book), the SelectMAP interface, for daisy-chaining the configuration process of multiple FPGAs, configuration loading from PROMs or compact flash cards, microcontroller-based configuration, an internal configuration access port (ICAP), and so on, depending on the specific family. A full explanation of these modes and techniques goes beyond the scope of this book. Readers should consult the user guide of the family of interest to find out more details about the various configuration possibilities.

Whatever the chosen configuration interface and mode, a bitstream file is a necessary prerequisite for the successful operation of any Xilinx FPGA. In every FPGA a *configuration logic* is built on the chip, with the purpose of implementing the different interfaces for exchanging configuration data and to interpret the bitstream to configure the device. A set of *configuration registers* defines the state of this configuration logic at a given moment in time. Configuration registers are the memory where the bitstream file has direct access. Actual configuration data are first written by the bitstream into these registers and then copied by the configuration logic on the configuration SRAMs. The details explained in the following section may vary in some minor aspects across different FPGA families, but the basic principles of configuration and bitstream composition hold true for all of them, that is, the bitstream composition and the underlying mechanisms in the configuration logic have been kept the same across different technological families.

The reference documents for understanding the configuration details of the FPGA families considered in this book are [191] for the Spartan 3 family, [218] for the Virtex II Pro family, and [100] for the Virtex 4 family.

*Bitstream files have the .bit extension.

2.2.1 Overall Bitstream Structure

A common full bitstream file is generally made up of five components:

- a comment part, stored mainly as the ASCII representation of the comment, reporting file name, creation date and time, and part model number.
- a dummy word (0xFFFFFFFF) and a synchronization word (0xAA995566), with the purpose of aligning the parser on 32-bit word boundaries and signaling the start of the actual configuration commands.
- a series of packets, with the respective headers and payloads, that are used to perform writes to the configuration registers of the configuration hardware of the device.
- a series of configuration words.
- another sequence of packets to write to the configuration register and issue start-up commands.

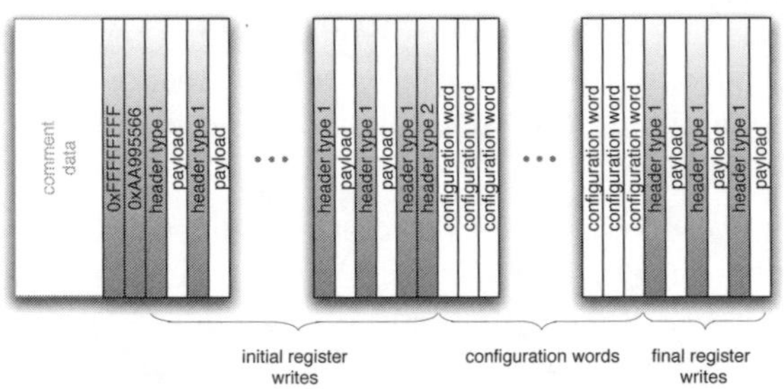

Figure 2.3: The general structure of a full configuration bitstream.

After the configuration logic has been aligned with the special synchronization word, the bitstream is processed and interpreted on 32-bit boundaries, word after word. The register writes at the beginning and at the end of the actual bitstream, i.e., the bitstream part after the sync word is realized via a packet structure of a header and a certain number of payload words.

2.2.2 Packet Headers

The header word defines the type of operation to perform on the configuration register, the address of the register to write to, and the number of subsequent data words destined to be written on that register. More details regarding the composition of the 32-bit type 1 header can be found in [100].

Type 1 packets are the most common packet type in the configuration bitstream. Type 2 packets are only used when the number of bits in the word count field of the type 1 header is insufficient to count a big number of configuration words. The type 2 packet has been created only for the purpose of reducing the overhead caused by the insertion of too many type 1 packets in long writes to the same configuration register. The header for the type 2 packet is defined in in [100]. The majority of the header is used to store the count of the subsequent words and no space is devoted to the register address, which is always defined in the previous word, which must be a type 1 packet header with word count set to zero. Therefore a type 2 packet must always follow a type 1 header, and it can be seen only as an extended word count field for the type 1 packet.

The significant data to understand the bitstream format are therefore in the header of type 1 packets. In these headers, the opcode field is usually set to a null operation or to a write operation in configuration bitstream. When in write mode, the last 5 bits of the Address field select one of the possible configuration registers, while the word count field indicates how many words the configuration logic must expect as a payload of the header.

2.2.3 Configuration Registers

The write operation performed by type 1 and type 2 packets have the purpose of writing to specific configuration registers. The registers of a Virtex 4 FPGA are listed in in [100]. Each of these registers controls a particular feature of the device. For the purpose of understanding the configuration process, the relevant registers are the CRC, IDCODE, FAR, and FDRI registers. All the other registers are used for controlling aspects other than the configuration. For more details readers should consult the documentation of the family of interest.

The **CRC** (Cyclic Redundancy Check) register is used to check the correct transferring of the configuration data by storing the result of a standard 16-bit checksum algorithm. The bitstream has a precalculated checksum value at the end of configuration words that is checked against the value calculated by the configuration logic. If the two values don't match an error situation is created and the configuration process must be repeated. The CRC verification can be disabled for each family by setting the appropriate bit of the COR register.

To avoid the downloading of a bitstream created for the wrong device, the **IDCODE** register must be written with a 32-bit word that is unique for every device. This word is also hard-wired into the configuration logic, and the comparison of the contents of the register with the hard-wired value can determine the correctness of the bitstream.

The *FAR* (Frame Address Register) is where the position of the first configuration frame is written, and it is thus important for knowing the physical location in partial bitstreams. In full bitstreams the value written to this register is always zero. This register is automatically incremented by the FPGA's configuration logic when the words that make up a configuration frame have been received, in such a way that only the first FAR value must be present in the bitstream file. The FAR address register is broken up into different fields, depending on the specific FPGA family taken into consideration†. The terminology used in the manufacturer's user guides has been updated to be uniform across every family. The **Block Type** field of the FAR address indicates broadly what kind of resources the frame with that value will configure. Different types are allowed: configuration of CLBs, IOBs, and CLOCK lines has the first block type address, then BRAM values and BRAM interconnect lines have their own block type values.

The **Major address** field indicates what column of resources the frame is currently configuring and the **Minor address** field indicates a frame inside a particular configuration column. The additional fields in the Virtex 4 FAR address have been introduced to address portions of configuration resources other than entire vertical columns of the device array. In particular, the **Top-Bottom** bit addresses the top or bottom half of the device and the **Row Address** is used to refer to a particular row of frames, in the bottom or top half, depending on the chosen model. Figure 2.4 aggregates all the possible information that is contained in a generic FAR word, either for the Virtex 4 family and for the Virtex II Pro and Spartan families, showing what is addressed in each field; for the families that do not support a bi-dimensional granularity in configuration, only one row of columns must be considered in the figure, while the whole addressing scheme, inclusive of the Row and Top/Bottom fields, holds for the Virtex 4 family.

The fields of the FAR address thus divide hierarchically the configuration memory space. The top level component is a row of resources, addressed by the Top/Bottom bit and by the Row address, respectively. Spartan 3 and Virtex II Pro can be considered as FPGAs having only one row of resources and thus adhere to this very same model. The next level in the hierarchy is defined by the Block Address together with the Major Address‡. These two fields define together which column in a row is configured. A frame within a column is then defined by its Minor Address, and there is no way to address configuration words within a single frame. The atomic configuration unit in a Xilinx FPGA is thus represented by the frame, as the addressing hierarchy clearly shows.

The **FDRI** register is used to write the words that make up a frame. The FPGA configuration logic implements the FDRI register as a shift register, so that a frame is configured while the next one is being shifted in. For this

†The Spartan 3 and Virtex II Pro FAR format are defined in [191] and [218], while Virtex 4 FAR format is defined in [100].

‡Xilinx documentation has various names for the Block Address: it is also called Column Address or Block Type field.

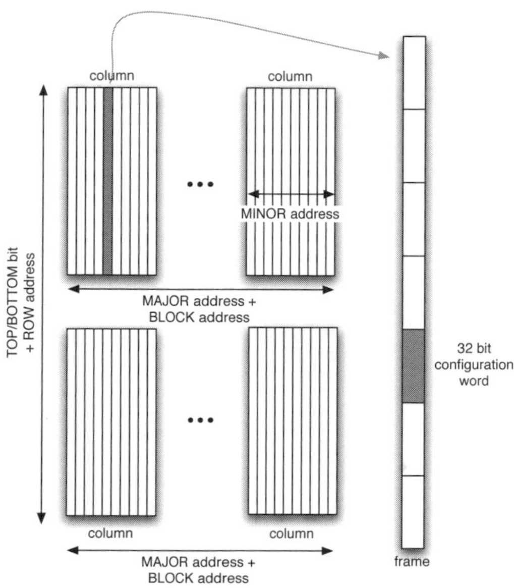

Figure 2.4: FAR addressing in Xilinx FPGAs.

reason in the bitstream there is some padding data in order to make the final writes to the configuration memory. As stated previously, the FAR address is automatically incremented whenever a whole frame has been written to the FDRI register.

The majority of a bitstream is thus made of the words that are written to the FDRI register, as shown in Figure 2.3.

2.2.4 Frame Indexing

The bitstream file structure, as seen in the previous paragraphs, contains a large write to the FDRI register with a sequence of 32-bit words that configure portions of the FPGA, written in a sequence that does not report the particular address (the FAR address) in the configuration memory to which the data are written. It is thus impossible to read a plain bitstream file and infer the portion of resources that it configures without the exact knowledge of the memory mapping of the particular FPGA to which the bitstream is going to be applied. In other words, there is the need to understand the logic with which the FAR address is incremented in the internal configuration machine of the FPGA in order to correctly map each frame read in the bitstream with the correct resources. The information provided in the vendor documentation hints at the addressing scheme used in the devices of the Virtex II Pro and Spartan families, but these details are not found in the documentation for the Virtex 4 family of devices. Figure 2.5, as an example of the information provided, reports the configuration memory mapping explained in the user guide for Virtex II Pro FPGAs. The *column type* field in

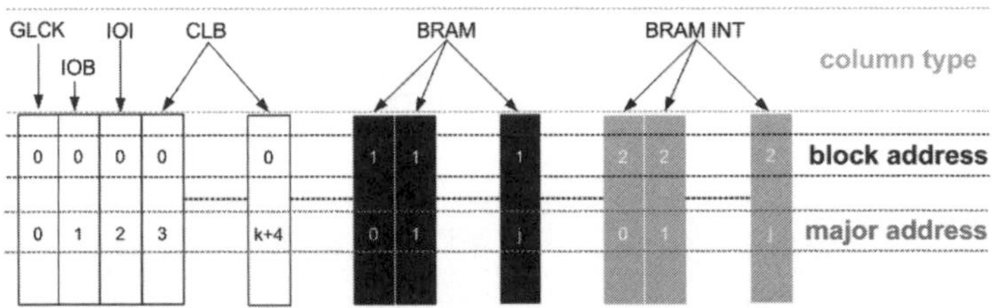

Figure 2.5: Virtex II Pro FPGA configuration memory mapping, from Xilinx Virtex II Pro User Guide [218].

the figure gives information about what particular column of resources in the array is being configured with the frames belonging to a column, and shows the progression of the configured resources as the FAR is automatically incremented by the configuration logic: the first column configures the central clock routing resources, then the next column the left IOB column, then their interconnections, then the columns of CLBs, and so on. A change from 0 to 1 in the *Block Address* field means that the frames of that particular column are configuring the contents of the block ram columns on the device, and a value of 2 in the *Block Address* means that the frames are configuring the interconnections of the ram cells with the rest of the array. The variables in this addressing scheme are given by the number of CLB columns and by the number of BRAM columns, marked with the letters k and j, respectively, in Figure 2.5.

2.3 FPGA Families and Models

The FPGAs used in this book belong to three distinct families, with different architectures and features. Table 2.1 lists the FPGA model(s) used to explore each of the three families used in this book. Since this book will mostly deal with low-level details of the architecture of the aforementioned FPGA chips, the following subsections will be devoted to illustrate the architecture of the devices in Table 2.1, by showing an internal representation of the resources of each model. As pointed out in Section 2.1.1, the general structure of the internals of an FPGA is a sea of logic gates, grouped in slices and CLBs,

Family	Model
Spartan 3 [217]	XCS200
Virtex II Pro [216]	XCVP7 XCVP20
Virtex 4 [219]	XC4VFX12

Table 2.1: FPGA families, models, and reference manuals used in this book.

into which heterogeneous hard-cores are embedded. This structure will be reflected in the following architectural description of the four models.

2.3.1 Spartan 3

The Spartan 3 family is the simplest of the three, and it offers fairly low cost devices. The available resources besides CLBs, block ram, and IOBs are dedicated 18-bit X 18-bit hardware multipliers and digital clock managers (DCM) hard-cores. In particular the chosen model, **XC3S200**, has a CLB array composed of 24 rows by 20 columns, interleaved by two columns in which reside 216 Kbits of block ram and 12 hardware multipliers. Four digital clock manager hard-cores are placed on the perimeter of the device in place of IOBs in the position corresponding to BRAM columns. An internal representation of the device is illustrated in Figure 2.6, where the resources that break the regular structure of the slice array have been highlighted in dark gray. Configuration of the resources is performed on a columnar basis, i.e., a subset of slices in a vertical column cannot be addressed for reconfiguration without configuring the remainder of the column as well.

2.3.2 Virtex II Pro

Virtex II Pro architecture provides more resources and an improved logic count with respect to the previous family. The two models of this family that have been studied in this work, **XCVP7** and **XCVP20**, have hard-core processor(s) on board, in order to exploit the speed and flexibility offered by a general purpose processor in standalone designs. In this way the performance in software execution of an external processor can be achieved directly within the same die of the FPGA. The processor model is the Power-PC 405. A representation of the internals of the two models is given in Figure 2.7, with resources other than CLBs and IOBs highlighted in dark gray as before. The available resources are an array of 4,928 and 9,280 slices, 44 and 88 for both 18Kb BRAMS and 18-bit x 18-bit multipliers and 4 and 8 DCMs for the two devices, respectively. Configuration is still performed on a per-column basis as in the Spartan 3 family.

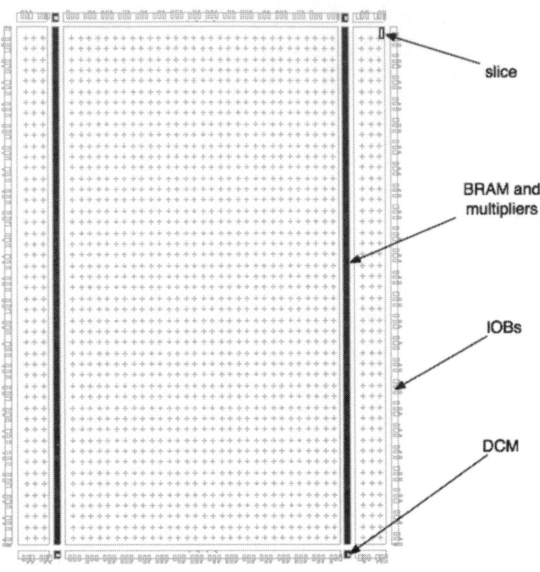

Figure 2.6: XCS200 internals as shown in XIlinx Floorplanner.

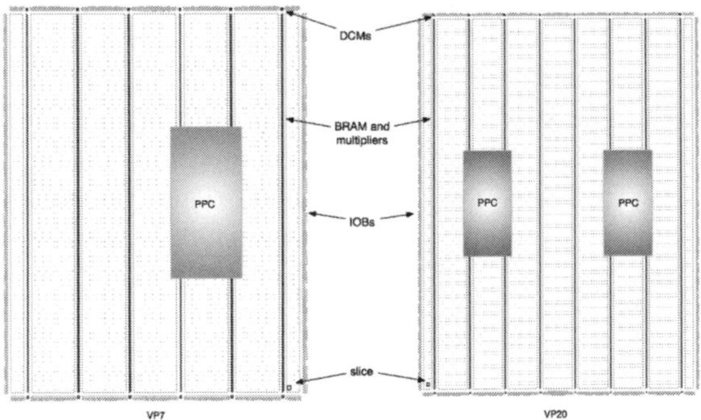

Figure 2.7: XCVP7 (left) and XCVP20 (right) internals as shown in Xilinx Floorplanner.

2.3.3 Virtex 4

The Virtex 4 family is the most recent of the families taken into consideration. The most important feature regarding dynamic reconfiguration is its capability to be configured without the columnar constraint. Configuration frames are now provided with a row address that points to a specific row of resources on the device. The single row of CLBs is not addressable individually, as the rows addressed by the frames are groups of CLB rows. This fact, however, makes a 2D reconfiguration scheme feasible. The quantity of hardware embedded on the silicon die has been improved, giving way to an even more heterogeneous array. The number of CLBs of the **XC4VFX12** is 64 by 24. The device internals are shown in Figure 2.8.

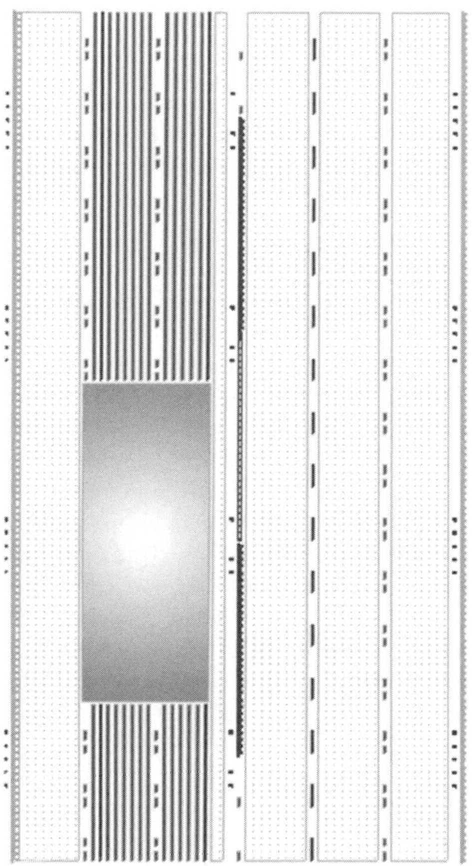

Figure 2.8: XCVFX12 internals as shown in Xilinx Floorplanner.

2.4 Configuration Conventions and File Formats

The following subsections will illustrate the conventions used by the vendor in numbering and addressing the resources on the array. These conventions are used extensively throughout the ISE[§] tools and some understanding of them is needed in order to correctly follow the steps in the flows for partial reconfiguration of the devices, namely for constraining pieces of logic inside particular regions. Both of the presented numbering schemes are visible by accessing the representation of a FPGA in FPGA editor. The two file types on which the analysis of this work has been performed will then be introduced.

2.4.1 Configuration Resources Numbering Scheme

The heterogeneous nature of the FPGA array requires the adoption of a numbering scheme to address the single resources in design tools. The model chosen by Xilinx is that of a separate numbering scheme for each resource category. Therefore there will be distinct numbering schemes for CLBs, BRAMs, multipliers, and so on. The process of numbering the resources is equal for all the types and it is location based: the resource with coordinates X0Y0 is always the one in the lower left corner of the device. The numbering of the subsequent resources in the array is determined just as if they were points in a bi-dimensional cartesian coordinate system. Having the resources numbered separately according to their type permits the correct identification of a particular element inside the set of elements of the same type, but this system does not allow the identification of the location of a particular resource on the device unless the exact architectural realization of the array is known.

An example of this numbering scheme is shown in Figure 2.9 for a generic type of resource. As the image shows, the first resource of the numbering scheme is placed in the lowest left corner, while the coordinates of other resources are incremented going to the right for X coordinates and up for Y coordinates. The maximum value of X is determined by counting the resources in each row and taking the maximum minus one. Analogously the maximum Y coordinate is given by taking the maximum among the column element numbers and subtracting one. Another important thing to note from Figure 2.9 is that *gaps* are allowed in the numbering system, so that, in the example, there will be no such resource as X2Y3 or X3Y3. In place of gaps there could be either another configurable resource, or another, non-configurable FPGA element.

[§]Integrated Software Environment, the set of software tools developed by Xilinx for going from an HDL description of the hardware to the bitstream(s) to be configured on the FPGA device.

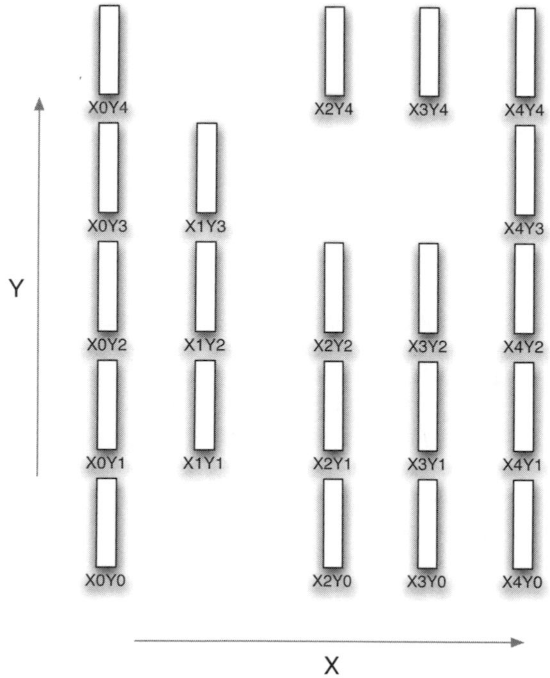

Figure 2.9: An example of the resource numbering system used to refer to resources across Xilinx devices.

2.4.2 The RPM Grid

In addition to the normal numbering scheme described in Section 2.4.1, a global coordinate system has been envisioned by the device manufacturer: in [198] this system is described as a reference for the realization of heterogeneous hardware macros. Hardware macros are pre-routed and pre-mapped hardware functionalities that can be included in a design. For the placement of the macros on the FPGA, the relative position of a resource with respect to another, independently of the resource type, must be known. For this reason a *global* numbering scheme has been introduced, called the RPM grid¶.

In essence the RPM grid is an array of positions that can be occupied by a particular resource. The position slots of this array are numbered in the exact same way as the numbering of single resource types, so that the exact resource occupation of a relatively placed hardware macro can be known in all of the resource types, and the relative positioning of those resources can be known. It is thus possible to place the macro in a new position on the FPGA,

¶RPM stands for Relatively Placed Macro.

given that that position has the same features of the original position where the macro has been developed, namely resources and inter-resource relative placements. This feature of the RPM grid will be useful during a partial bitstream relocation, where it is important to know both the resources that the bitstream configures and their relative placement on the FPGA die.

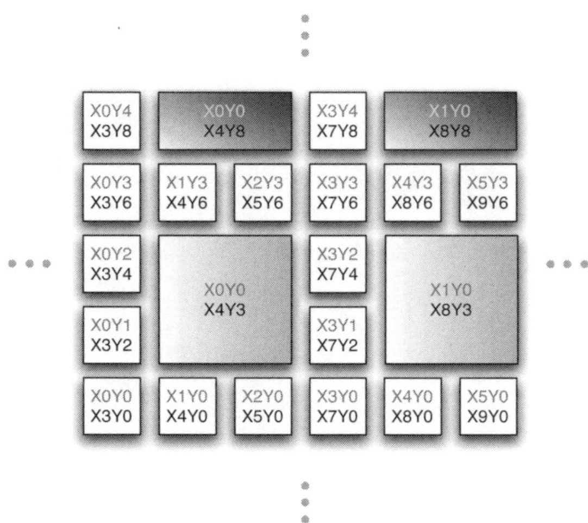

Figure 2.10: An example of the numbering style adopted in the RPM grid.

Figure 2.10 shows a small example that illustrates the pairing of the conventional numbering system described in Section 2.4.1 and the RPM numbering system described here. Different box sizes represent different kinds of resources, as in real devices where resources can be visually identified as belonging to a particular class according to their shape. In each box, the first line of text represents the coordinates in the conventional numbering system, while the second line has the coordinates of a sample RPM system. The point of reference of a resource in the RPM grid is its lower left corner, and the main constraint is that if the horizontal and/or vertical position of this point in the global visual representation is smaller than the position of the lowest left corner of another arbitrary resource, the former resource will have an RPM X coordinate and/or an RPM Y coordinate smaller than the latter. Increments in RPM X or Y coordinates are not always equal to one, as in the conventional numbering system, but can be other greater integer numbers, provided that the positional order of resources is respected as before.

2.4.3 UCF File Format

The UCF file format is used by the Xilinx toolchain to express some constraints on the design in order to guide the processes of placement, routing and mapping of the HDL descriptions of the soft-cores given by the developer. What is relevant for the analysis of the dynamic reconfiguration methodologies is the AREA GROUP class of constraints, which allows the definition of a *Reconfigurable Region*(RR[‖]) on the FPGA. The PAR (Place and Route) program will then read these constraints and reduce the configuration resource space accordingly. In the following excerpt from one of such UCF files a constraint about a reconfigurable region (or partial reconfigurable module, according to the terminology used by Xilinx) is defined.

```
INST "addsub_B" AREA_GROUP = "AG_PRMB";
AREA_GROUP "AG_PRMB" RANGE = SLICE_X28Y72:SLICE_X41Y127;
AREA_GROUP "AG_PRMB" RANGE = DSP48_X0Y26:DSP48_X0Y27;
AREA_GROUP "AG_PRMB" MODE  = RECONFIG;
```

The first line associates a component instance to a reconfigurable region. Then the physical constraints of the resources of that particular area are given, using the numbering scheme explained in 2.4.1. The second line defines the space of the region in the slice space, while the third defines the constraint in the space of the DSP48 resources. The last line, due to the *RECONFIG* word, states that the region defined will be affected by the reconfiguration process.

2.5 A Bird's-Eye View of Reconfigurable Systems

The versatility and reprogrammability of FPGAs comes at a price. Only a few years ago, the algorithms that could be implemented in a single FPGA chip were fairly small. In 1995, for example, the largest FPGAs could be programmed for circuits of about 15000 logic gates at most. Since a fast 32-bit adder requires a couple of hundred gates, the capabilities of such devices were somewhat restricted. More recently, though, FPGAs have reached a size where it is possible to implement reasonable sub-pieces of an application in a single FPGA part. The incorporation of reconfigurable array logic into a microprocessor provides an alternative growth path that allows application specialization while benefiting from the full effects of commoditization. Like modern reconfigurable logic arrays, a single microprocessor design can be employed in a wide variety of applications. Application acceleration and system

[‖] A Reconfigurable Region, or RR, is a portion of the device destined to be reconfigured with different functionalities. It is the place where the actual partial dynamic reconfiguration process takes place.

adaptation can be achieved by specializing the reconfigurable logic in the target system or application [53]. This has led to a new concept for computing: if a processor were to include one or more FPGA-like devices, it could in theory support a specialized application specific circuit for each program, or even for each stage of a program's execution.

The unlimited reconfigurability of an FPGA permits a continuous sequence of custom circuits to be employed, each optimized for the task of the moment. In the simplest scenario, which can be termed Compile Time Reconfiguration (CTR), the configuration of the FPGA is loaded at the end of the design phase, and it remains the same throughout the whole time the application is running. In order to change the configuration one has to stop the computation, reconfigure the chip resetting it, and then start the new application. CTR was for some years the only kind of reconfiguration available for FPGAs. With the evolution of technology, though, it became possible to considerably reduce the time needed for the chip reconfiguration: this made it possible to reconfigure the FPGA *between* different stages of its computation since the induced time overhead could be considered acceptable. This process is called Run Time Reconfiguration (RTR), and the FPGA is said to be *dynamically reconfigurable*. RTR can be exploited by creating what has been termed *virtual hardware* [125, 69] in analogy with the concept of *virtual memory* in general computers. Consider, for instance, an application that is too big to fit into a particular FPGA: one can *partition* it into n smaller tasks, each one fitting on the chip. Then it is possible to load task 1 on the chip, execute it, then reconfigure the FPGA for task 2 and execute it, and so on until task n is finished. This idea is called *time partitioning*, and has been studied extensively in the literature (see [34, 37, 114, 33]).

A further improvement in FPGA technology allows modern boards to reconfigure only *some* of the logic gates, leaving the other ones unchanged. This *partial reconfiguration* is of course much faster in case only a small part of the FPGA logic needs to be changed. When both these features are available, the FPGA is called *partially dynamically reconfigurable*. Although there are several techniques to exploit partial reconfiguration (e.g., [110, 95, 98]), there are only a few approaches for frameworks and tools (e.g., [95, 98, 30, 165, 86]) to design dynamically reconfigurable SoPC (e.g., [24, 58, 68]). Examples of such frameworks are the operating systems for reconfigurable embedded platforms which have been analyzed in [178]. In [192] authors have presented a run-time system for dynamical on-demand reconfiguration. Several research groups [185, 82, 11, 74, 123, 110, 117, 108] have built reconfigurable computing engines to obtain high application performance at low cost by specializing the computing engine to the computation task; some preliminary results can be found in the literature [31, 87, 171, 184, 85, 74], but no general framework and no publicly available tools are, to the best of our knowledge, available.

Due to the capabilities described above, FPGAs can be used to create hardware/software platforms that keep their flexibility at runtime, allowing the development of SoPCs. Modern FPGAs can also contain a general-purpose processor, which can be both a physical CPU embedded in the FPGA fabric, or a soft-core, mapped to a part of the FPGA. In both scenarios there is a software (SW) application running on the processor (or multiple processors) which realizes some of the system functionalities, usually acting also as a controller for the hardware (HW) components and interfacing with them. The SW part of a reconfigurable system can be either a standalone code, dealing directly with HW at a low level, or a complete operating system, including multiprocessing and resource scheduling.

A standalone code is usually an application that uses SW libraries exporting functions to interface with HW components. This approach can be acceptable for small systems, involving few components and configurations, but as soon as the complexity of the system increases, it becomes more difficult to develop a complete application dealing with all those aspects. The use of an operating system allows more flexibility on both sides, since it is possible to implement the SW part as one or multiple *userspace* processes, introducing complex inter-process communication systems and scheduling techniques. HW management is remitted to the OS, offering the processes an interface to access system peripherals at a higher level of abstraction. The counterpart for this added flexibility is the necessity of adding support to a standard operating system for reconfiguration-specific HW and for reconfigurable components. This means that the HW and SW parts of the system must be designed in order to allow the creation of a reconfigurable architecture. More details regarding the operating system support for the dynamic reconfiguration will be provided in Chapter 4.

2.6 Reconfiguration Characterization

The reconfiguration capabilities of FPGAs give the designers extended flexibility in terms of hardware maintainability. As will be shown in Section 2.1, FPGAs can change the hardware functionalities mapped on them by taking the application offline, downloading a new configuration on the FPGA (and possibly new software for the processor, if any) and rebooting the system. Reconfiguration in this case is a process independent of the execution of the application. The process is called **Static Reconfiguration**.

A different approach is the one that considers reconfiguration of the FPGA as part of the application itself, giving it the capability of adapting the hardware configured on the chip resources according to the needs of a particular situation during the execution time. This particular case is called **Dynamic Reconfiguration**. In the case of dynamic reconfiguration, therefore, the reconfiguration process is seen as part of the application execution, not as a stage prior to it.

2.6.1 Complete vs. Partial Reconfiguration: An Overview of Different Techniques

The easiest way in which an FPGA can be reconfigured is called *complete.* In this case the configuration bitstream, containing the FPGA configuration data, provides information regarding the complete chip and it configures the entire FPGA. That is why this technique is called complete. With this approach there are no particular constraints that have to be taken into account during the reconfiguration action. Obviously, that does not mean that the designer is allowed whatever he/she wants only because she/he is using a configuration technique based on a *complete-reconfiguration.* In fact, if two different bitstreams implement two functionalities that have to work one after the other (see Figure 2.11 for an example of such a scenario), the designer has to take into account where to store the data between these two configurations.

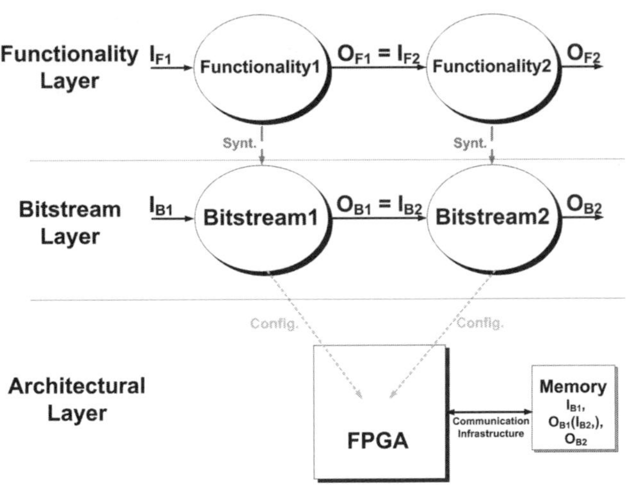

Figure 2.11: Communication problem between two different configurations.

The main disadvantage of an approach based on the complete reconfiguration technique is the overhead introduced into the computation by the reconfiguration. In order to cope with this situation a *partial reconfiguration* approach has been proposed. Partial reconfiguration is useful for applications that require the load of different designs into the same area of the device or the flexibility to change portions of a design without having to either reset or completely reconfigure the entire device [59]. For current FPGA devices, data are loaded on a column basis, with the smallest load unit being a configuration bitstream *frame*, which varies in size based on the target device. Active partial reconfiguration of Virtex devices, or simply *partial reconfiguration*, is accomplished in either slave SelectMAP mode or Boundary Scan, JTAG mode. Instead of resetting the device and performing a complete reconfiguration, new data are loaded to reconfigure a specific area of the device, while the rest of the device is still in operation [96].

The scenario shown in Figure 2.11 turns into the scenario proposed in Figure 2.12.

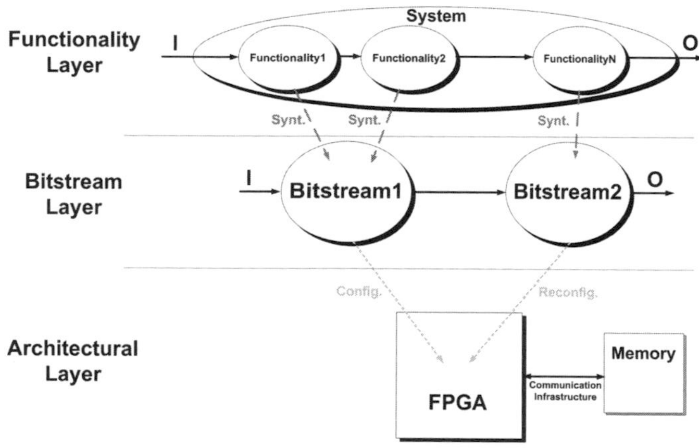

Figure 2.12: Partial reconfiguration scenario.

Using an approach based on partial reconfiguration, such as the one proposed in Figure 2.12, the basic idea is to partition the system in a set of functionalities $f_1, f_2, \ldots, f_n$ able to produce a set of bitstreams $b_1, b_2, \ldots, b_n$ that are not used to reconfigure the entire system but just a known portion of it. The first bitstream is obviously a complete bitstream but the other functionalities are downloaded to reconfigure just portions of the architecture, as proposed in Figure 2.13.

With such a scenario the reconfiguration time of a portion of the FPGA is hidden by the computation of the remaining part. According to this last statement it is easy to see that an important component is still missing in the model proposed in Figures 2.12 and 2.13. In order to be able to hide the reconfiguration time it is not only necessary to partition the FPGA to obtain the ability to compute partial reconfiguration bitstream, but it is also necessary to guarantee that a reconfiguration is not going to imply a standby in the computation of the not-involved logic of the FPGA. Such a scenario

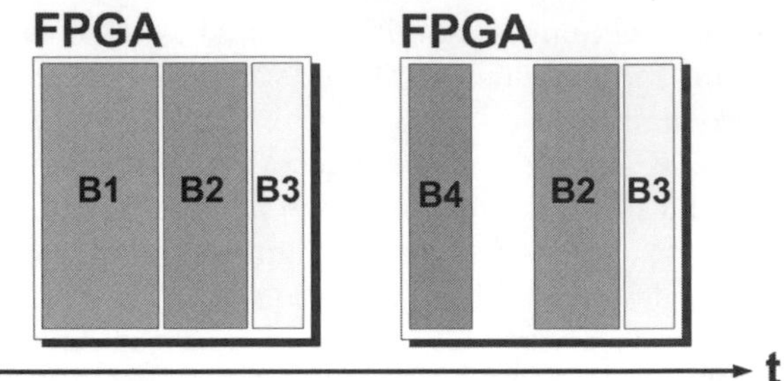

Figure 2.13: Partial reconfiguration example.

helps define *Dynamic Partial Reconfiguration*, which is performed when the device is active. Except during some inter-design communication, certain areas of the device can be reconfigured while other areas remain operational and unaffected by the reprogramming.

2.6.2 Reconfigurable Architectures: The Five Ws

FPGA configuration capabilities allow a great flexibility in HW design and, as a consequence, they make it possible to create a vast number of different reconfigurable systems. These can vary from systems composed of custom boards with FPGAs, often connected to a standard PC or workstation, to standalone systems including reconfigurable logic (RL) and general purpose processors (GPP), to System-on-Chip's, completely implemented within a single FPGA mounted on a board, with only few physical components for I/O interfacing. There are different models of reconfiguration, which can be classified according to the following scheme (an extension of the one proposed in [209]):

- *who* controls the reconfiguration;
- *where* the reconfiguration controller** is located;
- *when* the configurations are generated;
- *which* is the granularity of the reconfiguration;
- in *what* dimension the reconfiguration operates.

**The element that is responsible for the physical implementation of a reconfiguration process, i.e., in Xilinx FPGA the ICAP controller.

2.6.2.1 Who and Where

The first subdivision (*who* and *where*) is between *external* and *internal reconfiguration. External reconfiguration* implies that an active array may be partially reconfigured by an external device such as a personal computer, while ensuring the correct operation of those active circuits that are not being changed.

Internal or self-reconfiguration, on the other hand, is performed completely within the FPGA boundaries. Clearly the integrity of the control circuits must be guaranteed during reconfiguration, so by definition self-reconfiguration is a specialized form of dynamic reconfiguration [183]. An important feature in FPGA architectures is the ability to reconfigure not only all the device but also a portion of it while the remainder of the design is still operational. Once initially configured, self-reconfiguration requires an internal reconfiguration interface that can be driven by the logic configured on the array. Starting with Xilinx Virtex II parts, this interface is called the internal configuration access port, ICAP [60, 97]. These devices can be configured by loading application-specific data into the configuration memory, which is segmented into frames, the smallest unit of reconfiguration. The number of frames and the bits per frame are different for the different devices of the Virtex II family. The number of frames is proportional to the CLB width of the device. The number of bits per frame is proportional to the CLB height of the device.

2.6.2.2 When the Configurations are Generated

The generation of the configurations (*when*) can be done in a completely *static way* (at design time) by determining all the possible configurations of the system. Each module must be synthesized and all possible connections between modules and the rest of the system must be considered. Other possibilities are *run-time placement of pre-synthesized modules*, which requires dynamic routing of interconnection signal, or completely *dynamic modules generation*. This last option is currently impracticable, since it would require run-time synthesis of modules from VHDL (or other HW description language) code, a process requiring prohibitive times in an online environment.

2.6.2.3 Which is the Granularity of the Reconfiguration?

Reconfiguration can take place at very different granularity levels (*which*), depending on the size of the reconfigured area. Two typical approaches are *smallbits* and *module based*. The first type consists of modifying a single portion of the design, such as single configurable logic blocks (CLB) or I/O blocks parameters (as described in [60]), while the second type involves the modification of a larger FPGA area by creating HW components (modules) that can be added and removed from the system. Each time a reconfiguration is applied, one or more modules are linked or unlinked from the system.

2.6.2.4 In What Dimension the Reconfiguration Operates: The Mono-Dimensional and the Bi-Dimensional Scenarios

The last property is the *dimension*. One can distinguish between two different possibilities: mono-dimensional (1D) and bi-dimensional (2D) reconfiguration. In **1D reconfiguration** the resources affected by the process always span the whole height of the device. In other words, it is impossible to configure one set of resources without having to reconfigure the whole columns they span as well. This reconfiguration scheme is imposed by the architecture of some devices such as those in Xilinx Spartan 3 and Virtex II families because the smallest addressable configurable unit in these devices is the frame, a vertical line of resources. On the other hand, **2D reconfiguration** allows the partitioning of the configurable resources in a 2D space, allowing the definition of different reconfigurable regions one on top of the other.

The 1D limitation has been overcome in more recent FPGA families such as Virtex 4 and Virtex 5, which can reconfigure specific portions of the device without affecting the entire columns of corresponding resources. In this way more sophisticated architectures can be designed, exploiting the added dimension to the freedom in the choice of reconfigurable areas. Figure 2.14 illustrates these two different schemes. The 1D scheme is presented in Figure 2.14(a), showing how the whole FPGA height is occupied by a reconfigurable region. In such a scenario it is not possible to define two reconfigurable regions that share two different parts of the same column, so it is compulsory to assign a column to a single reconfigurable region (see Figure 2.14(a)): this observation explains why, even using a 2D placement, it is not possible to perform a true 2D reconfiguration.

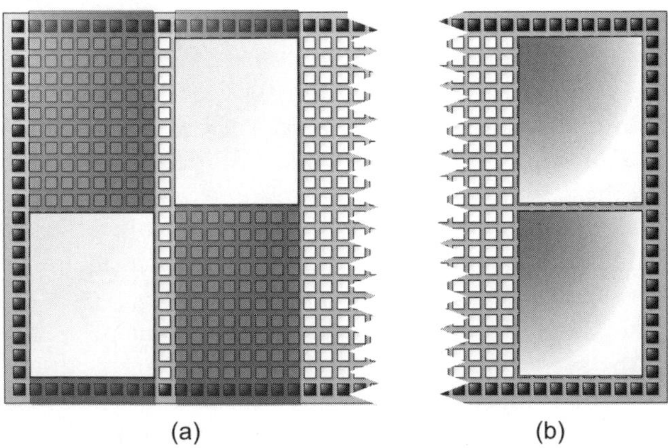

Figure 2.14: (a) 1D and 2D placement constraints vs. (b) 1D and 2D reconfiguration.

On the other hand, in Figure 2.14(b), the constraint on vertical occupation is relaxed, thus the reconfigurable areas do not span the entire height of the device anymore. In this context it is now possible to reconfigure regions in both a 1D and a 2D fashion, thus providing a greater flexibility in choosing the best floorplan.

2.7 Reconfigurable Architecture Examples

The *PipeRench* architecture, proposed in [74], introduces the concept of hardware virtualization to provide the possibility of implementing a design of any size on a given configurable pipeline architecture of arbitrary capacity. The *PipeRench* system provides both the extremely high-speed reconfiguration necessary for hardware virtualization and compilation tools for this architecture. In this way it is possible to find a solution for both the problems that inhibit the deployment of applications based on run-time reconfiguration. The first is that design methodologies for partial reconfigurable applications are completely ad hoc, while the second one is the lack, in existing FPGAs, of reconfiguration mechanisms that adequately support local run-time reconfiguration. This solution is suitable for a scenario in which available resources are not enough for the computation, so it is possible to exploit the *reconfigurable pipeline* to virtualize pipeline stages. However, the *reconfigurable pipeline* structure introduces some relevant constraints that limit the freedom of the design.

The *MorphoSys* architecture, presented in [146], is a reconfigurable computer architecture targeted to computational intensive applications. It consists of a *TinyRISC* processor and a *RC-Array*, defining the reconfigurable hardware unit. The RC-Array is composed of a bi-dimensional array of reconfigurable cells (RC), whose configurations are stored in the *context memory*. During the execution, configuration data are fetched from the *context memory*, while computational data for the *RC-Array* are loaded in the *framebuffer* from external memory. The execution model of the *MorphoSys* processor is based on the partitioning of applications in sequential and data-parallel tasks; the first are executed by the TinyRISC processor, while the latter are mapped on the RC-Array. To use the *MorphoSys* architecture it is necessary to write both the *RC-Array* configuration program and the program for the *TinyRISC* processor. The first can be realized by using a specific assembler language, while the second one can be obtained from a C compiler. However, the current version of the compiler is not able to manage the *RC-Array*, so the control instructions have to be manually inserted by the programmer. Furthermore, it is not explained how it is possible to derive the sequential and parallel tasks from a given application and how they are managed by the scheduler.

In [160], [24] the hardware subsystem of the reconfiguration control infrastructure sits on the on–chip peripheral bus (OPB). The microprocessor, PowerPC or Microblaze, communicates with this peripheral over the OPB bus. The hardware peripheral is designed to provide a *lightweight* solution to reconfiguration. It employs a read/modify/write strategy. Only one frame of data is worked on at one time. In this way external memory is not needed to store a complete copy of the configuration memory. The program installed on the processor requests a specific frame, then the control logic of the peripheral uses the ICAP to do a readback and loads the configuration data into a dual-port block RAM. One block RAM can hold an XC2V8000 data frame easily. When the read-back is complete, the processor program directly modifies the configuration data stored in the BRAM. Finally, the ICAP is used to write the modified configuration data back to the device. The software subsystem is implemented using a layered approach. This solution allows a change in the implementation of the lower layers without affecting the upper layers, and proved useful for debugging. There are functions for downloading partial bitstreams stored in the external memory, for copying regions of configuration memory, and pasting it to a new location [160].

In [10] the authors proposed the *Splash* processor. Splash is a special-purpose parallel processor that is able to exploit temporal parallelism (pipelining) or data parallelism (single instruction multiple data stream) present in the applications. In this processor the computing elements are programmable FPGA devices. *Splash* boards consist of 16 FPGAs (X1. . . X16), a crossbar switch, and a seventeenth FPGA (X0), which acts as a control element for the board. Within a board, an FPGA is connected to its left and right neighbor and to the crossbar switch. The boards are connected to each other in a chain, and the X0 element of each board is also connected to the interface board. The workstation performs a wide range of operations. It acts as a general controller for the reconfiguration of FPGA elements and crossbar switches, sends computational data and control signals to the array, and collects the results. Dynamic reconfiguration in the *Splash* model consists of modifying two kind of elements within a board: the crossbar switch and/or the processing elements. In the first scenario, reconfiguration of the crossbar switch interconnections allows an easy way to modify the data flow in the system without the need to modify the single computing elements; in the second one, single FPGAs are reconfigured to change the computation performed on the data. Programming for the *Splash* system is done by writing the behavioral description of the algorithm using VHDL, which goes through a process of refinement and debugging using the *Splash* simulator. The algorithm is then manually partitioned on the different processing elements.

In [57] the authors proposed an architecture that is logically divided in two parts: a static side and a reconfigurable one. The static side contains a standard IBM CoreConnect technology [48] and two different buses: the OnChip Peripheral Bus (OPB) and the Processor Local Bus (PLB). Both these buses are implemented using a standard 32-bit wide data bus and 32-bit wide address bus and they are defined using the same working frequency. The communication between the fixed and the reconfigurable area is obtained exporting from the fixed side the necessary OPB signals used to define the correct bind of each reconfigurable module to it. The reconfigurable side is composed of one or more modules that can be dynamically replaced. The proposed architecture uses a mono-dimensional approach to reconfigure itself, thus reconfigurable modules are as high as the entire FPGA, according to the reconfiguration-by-column technique. Reconfigurable modules are attached to the OPB bus through signals exported by the fixed component and acts like a slave on the OPB. This architecture has two main weakness in the communication and the portability over different FPGA families. The communication infrastructure between the fixed side and the reconfigurable one implies that, while a reconfiguration process takes place, inter-module communications are disrupted. The proposed infrastructure also limits the number of reconfigurable modules. The second problem is caused by the instantiation of a Power-PC processor inside the fixed side. This limits the portability of the architecture to only FPGAs containing this processor inside their die.

Summary

This chapter presents low-level configuration details of the FPGA devices. It describes a particular technique called Partial Dynamic Reconfiguration that has been viable with most recent FPGA devices. This technique will be described in detail in Chapter 7. To fully understand what this technique is, the concepts of reconfigurable computing, static and dynamic reconfiguration, and the taxonomy of dynamic reconfiguration itself must be analyzed. In this way partial dynamic reconfiguration can be correctly placed in the set of system development techniques that are possible to implement on a modern FPGA chip. In order to better understand the details of the partial dynamic reconfiguration, part of this chapter is devoted to explore the file type that contains the configuration information to initialize the state of these reconfigurable chips, the bitstream file. To describe the bitstream composition, details have been aggregated from the three different Xilinx families of FPGAs examined in this book: Spartan-3, Virtex-II Pro, and Virtex 4.

Exercises and Problems

1. Define the term *bitstream* and describe the differences between bitstreams and *partial* bitstreams.
2. Is it possible for two partial bitstreams to have the same x-axis coordinate on a Xilinx Virtex II Pro device? What about a Virtex 4?
3. For what is the information saved in the FAR register useful?
4. Find and list as many architectural differences as you can among the three different FPGAs presented.
5. Differentiate between FDRO and FDRI.
6. List and describe as many differences as you can between complete and partial reconfiguration.

3

Reconfigurable Hardware Design

To gain insight into the advantages of partial dynamic reconfiguration, let us introduce the following motivational example. Consider a simple FPGA with three reconfigurable units and an application represented by the task graph shown in Figure 3.1(a) (where a node with label n/m has a latency of n cycles and needs m reconfigurable units). The time partitioning approach

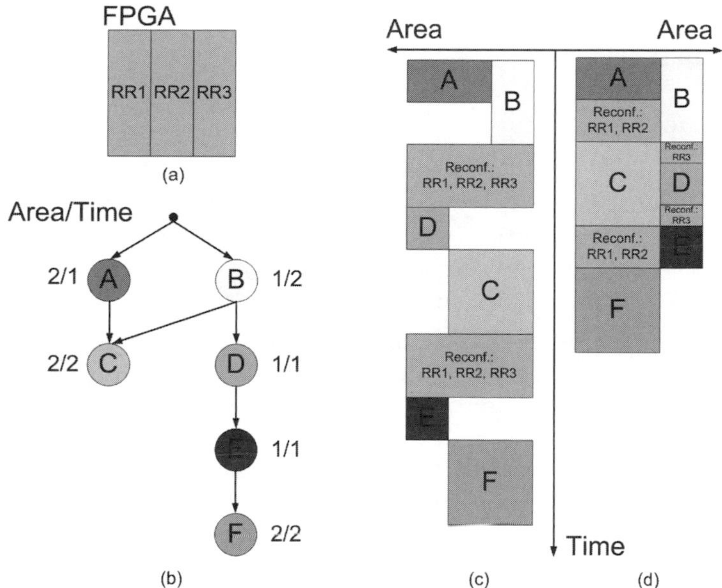

Figure 3.1: A reconfigurable device (a), a program specification (b), the time partitioning schedule (c), and the optimal schedule (d).

will at best identify the scheduling in Figure 3.1(b), since the chip, for the sake of simplicity, is considered an homogeneous device organized in three different reconfigurable regions, and must be reconfigured completely each time. The size of each task is represented horizontally, while time is shown increasing downwards. Gray areas represent reconfiguration time. If instead one allows the selective reconfiguration of certain areas of the FPGA corresponding to a task (or to groups of tasks), we can obtain the better schedule shown in Figure 3.1(c), which gives a much smaller total latency. The idea is to hide the latency due to the reprogramming of the logic gates by reconfiguring some parts of the chip while other parts are performing useful computations. As shown in this very simple example, not only does one gain parallelism among reconfiguration and execution, but it is also possible to expose more parallelism among the tasks themselves (tasks C and E). This allows a good exploitation of the capabilities of partial dynamic reconfiguration.

This chapter explains how to best exploit such capabilities, starting from a given specification, by defining specific solutions to the partitioning and the scheduling problem tailored for self-dynamically reconfigurable systems. The problem at hand will be described from the point of view of the actual physical architecture. Then a novel model will be used to describe it in a precise mathematical way to allow its analysis.

3.1 Model

In order to formally describe the properties of the actual FPGAs it is necessary to define a model that will allow us to clarify the problem at hand, and thus to better study its properties. We will strive to build a model that is general enough to be applicable to several different chips, but also not too abstract nor simple, thus retaining the important features of the problem.

The specification is given in terms of a Data Flow Graph (DFG) $\langle O, P\rangle$, where the node set O represents the *operations* and the arc set P represents the *precedence relations*: arc $o_i \rightarrow o_j$ indicates that operation $o_i \in O$ must terminate before operation $o_j \in O$ starts. For convenience, the DFG also includes the dummy start node and the dummy end node from the input graph $\langle O, P\rangle$ in graph $\langle \mathcal{S}, \mathcal{P}\rangle$ correspond to singleton subsets, respectively, S_e and S_s. In the setting of the DRESD project [152], the DFG is usually obtained semi-automatically from a high-level specification (SystemC, C, VHDL) [64].

The reconfigurable device is modeled as a set $U = \{u_1, u_2, \ldots, u_{|U|}\}$ of *reconfigurable units* (RU): each $u \in U$ is made up of ρ_u CLBs. The RUs are *linearly ordered*, by which we mean that u_k is adjacent to $u_{k\pm1}$ on the FPGA, for every $1 < k < |U|$. The operations can be physically implemented on the target device using a set E of *execution units* (EUs), which correspond to different configurations of the resources (RUs) available on the device.

In such a scenario, the reconfigurable partition and scheduling problem amounts to

1. partitioning the operations O into a collection of subsets;

2. mapping each subset S onto a compatible EU $e \in E$;
3. assigning the EU implementing each subset S defined in step 1 to a subset of RUs;
4. scheduling both the reconfiguration and the execution of each subset S according to a number of precedences and resource constraints.

Resource sharing occurs at EU level: different sets of operations may exploit the same RUs if they are executed in disjoint time intervals. Moreover, when they also share the same EU, they can be executed consecutively with a single reconfiguration at the beginning. Due to the interaction with the other constraints, even if resource sharing is possible, it is not *always* advisable.

Given a DFG $\langle O, P \rangle$, any partition of O yields a new DFG $\langle \mathcal{S}, \mathcal{P} \rangle$ with a node S for each subset in the partition and an arc $S \to S'$ for each pair of subsets such that $\exists o \in S, o' \in S' : o \to o' \in P$.* We consider as unfeasible all partitions producing non-acyclic graphs.

Given any subset of operations S and any of its feasible EU implementations e, we assume that suitable algorithms allow us to readily compute a latency $l_{e,S}$, a size r_e and a reconfiguration time d_e. The dummy subsets S_s and S_e have zero latency, size, and reconfiguration time.

Due to the previous remarks, it is possible to define as a solution of the problem a function that specifies for each subset of operations $S \subset O$: 1) whether it belongs to the partition ($x_S = 1$) or not ($x_S = 0$) and (in the former case) 2) onto which EU e_S it is mapped; 3) to which subset U_S of RUs it is assigned; 4) in which time step $\bar{t}_S$ its reconfiguration starts; 5) in which time step t_S its execution starts.

$$\sigma : 2^O \to \{0, 1\} \times E \times 2^U \times T \times T : S \mapsto (x_S, e_S, U_S, \bar{t}_S, t_S).$$

where T is the scheduling time horizon.

The following constraints must be defined:

- each EU must fit in the RUs assigned to it: $r_{e_S} \leq \sum_{u \in U_S} \rho_u$ for all S such that $x_S = 1$;
- the execution must follow the reconfiguration: $\bar{t}_S + d_S \leq t_S$ for all S such that $x_S = 1$;
- precedences must be respected: $t_S + l_S \leq t_{S'}$ for all S and S' such that $x_S = x_{S'} = 1$ and $S \to S' \in \mathcal{P}$;
- two sets of operations can not be running at the same time on the same RU: $t_S \leq t_{S'} < t_S + l_S \Rightarrow U_S \cap U_{S'} = \varnothing$ for all S and $S' \neq S$ such that $x_S = x_{S'} = 1$;

*The dummy start node and the dummy end node from the input graph $\langle O, P \rangle$ in graph $\langle \mathcal{S}, \mathcal{P} \rangle$ correspond to singleton subsets, respectively, S_e and S_s.

- while a set of operations is in execution, its RUs cannot be reconfigured: $t_S \leq \bar{t}_{S'} < t_S + l_S \Rightarrow U_S \cap U_{S'} = \varnothing$ for all S and $S' \neq S$ such that $x_S = x_{S'} = 1$;
- while a set of operations is in reconfiguration, no other set of operations can be reconfigured unless it uses completely different RUs or exactly the same RUs with exactly the same EU (in the latter case, one of the two reconfigurations is actually skipped) $\bar{t}_S \leq \bar{t}_{S'} < \bar{t}_S + d_S \Rightarrow U_S \cap U_{S'} = \varnothing$ or ($e_S = e_{S'}$ and $U_S = U_{S'}$) for all S and $S' \neq S$ such that $x_S = x_{S'} = 1$;
- no reconfiguration can occur on an RU between a reconfiguration and an execution: $\bar{t}_S \leq \bar{t}_{S'} < t_S \Rightarrow U_S \cap U_{S'} = \varnothing$ or ($e_S = e_{S'}$ and $U_S = U_{S'}$) for all S and $S' \neq S$ such that $x_S = x_{S'} = 1$;
- the RUs must be adjacent[†]: $\forall S, |U_S| > 1$, $\exists h, k \in \mathbb{N} : U_S = \{u_h, u_{h+1}, \ldots, u_{h+k}\}$.

The aim is to minimize latency:

$$\text{argmin}_\sigma \; t_{S_e}. \tag{3.1}$$

The total latency depends, among other things, on finding a *good* partition $\mathcal{S}$ of O. However, this task is *not* independent of mapping and scheduling: the three problems have to be solved together to achieve optimality.

This approach, however, is impractical because of the overwhelming size of the search space: for each of the $B_{|O|}$ possible partitions of O (where B_n is the n-th Bell number and grows super-exponentially with n), one must consider several possible implementations, $2^{|U|}$ possible locations, $|T|$ possible reconfiguration start times, and $|T|$ possible execution start times. In an ILP formulation it would involve, even for small specifications, a huge number of 0-1 variables.

One could try to find a partitioning $\mathcal{S}$ *independently* of the scheduling, but still trying to do it in such a way that the result is still *good enough* for scheduling purposes. As a particular case, one could simply ignore the partitioning process and schedule the original DFG. This amounts to choosing $\mathcal{S} = O$ in our model.

3.2 Partitioning for Reconfigurable Architectures

Different approaches to the exploitation of the reconfiguration capabilities of devices such as FPGAs have been explored in the literature, with notable ex-

[†]To minimize communication overhead.

amples in [150, 144, 113, 36, 35, 195]. Among these we can identify approaches that only carry out *temporal* partitioning of the device's resources, i.e., the whole device is reconfigured at once when the application needs to be, and approaches that also perform *spatial* partitioning of the device into reconfigurable units of smaller size. The latter approaches are those that are closest to our point of view, and are elsewhere referred to as *dynamically reconfigurable* architectures. Let us first focus on some of the works that propose total reconfiguration.

3.2.1 Temporal Partitioning Approaches

A quite straightforward approach is proposed in [150], in which the authors have the objective of partitioning a specification that would not fit on the FPGA as a whole into a set of *subprograms*, to be configured sequentially on the device. The proposed solution involves two steps:

- as a first step, an acyclic data flow graph is partitioned to produce a set of configurations, $\{c_1, c_2, \ldots, c_n\}$, each of which fits singularly on the FPGA;
- the partitions thus obtained are configured one after the other onto the device, and executed serially. Figure 3.2 shows the sequence of configurations loaded onto the device: all of the device is reconfigured for the execution of each partition.

Two approaches are proposed for the partitioning of the specification, one aiming at minimizing the temporal latency, the other trying to minimize the communication overhead generated by the partitioning. Both algorithms start out by making sure that the data dependencies encoded in the data flow graph are respected in the partition that will be generated, and they do so essentially by traversing the graph in a breadth-first fashion, and assigning a so-called ASAP level to each vertex v, defined as:

$$\lambda(v) = 1 + \max_{w \in FanIn(v)} \lambda(w)$$

The first approach simply orders the vertices by increasing the value of λ, and adds vertices to a partition until no more will fit on the device, at which point it creates another partition and continues similarly until all the specification has been considered. The second approach, on the other hand, aims at minimizing the overhead due to data exchange between different partitions by trying, when possible, to assign successors of a given node to the same partition.

An estimate of the total execution time was provided in this work as $T_{total} = (n \cdot T_{reconf}) + T_{exec}$, simply stating that the total time required for executing the partitioned specification is given by the total reconfiguration time needed added to the total execution time needed. The authors themselves, however,

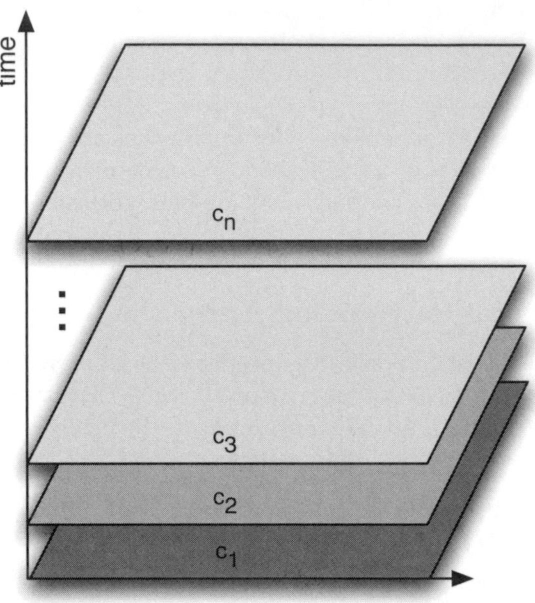

Figure 3.2: Temporal partitioning: the sequence of configurations.

noted how the reconfiguration time was, for their hardware devices, *orders of magnitude* larger than the execution time, and suggested improvements to tackle this disadvantage, such as applying the method to applications in which each configuration can be used multiple times before being swapped out for the next one, such as applying the first phase of a JPEG encoder to all the input images before going on to configuring the subsequent stages.

A similar approach is exposed in [94], which also considers some amount of spatial partitioning while still considering total reconfiguration of the physical devices due to the use of multiple FPGAs in a single system.

A series of papers [144, 113, 145, 112, 114, 70, 78] present approaches to the temporal partitioning of a system specification by exploring different ideas and improvements. In [112], an Integer Linear Programming (ILP) model is proposed to describe the temporal partitioning problem. It takes into account the dependencies between different parts of the specification and the resource availability on the FPGA, and also models the constraints due to memory needs for exchanging data between the different partitions identified. As its input, this approach uses a task graph, i.e., a graph in which each node is not a single operation, but a set of operations instead. This constrains the algorithm to miss a large number of the possible solutions, since the minimum working granularity is set at the task level. Also, many of the parameters needed by the model to function properly are to be supplied by the designer, apparently with no general guideline for choosing reasonable values, e.g., for the maximum number of partitions or the maximum number of functional units to be instantiated for each type.

The work proposed here is extended in [144] with the application of a genetic algorithm that adds spatial partitioning in order to split the specification onto multiple interconnected reconfigurable devices. The lack of indications towards the choice of the ILP model parameters is tackled in [113], which revises some aspects of the model itself: each task is associated with a set of possible implementations, each characterized with an estimate of its latency and area usage. The ILP solver chooses among them which to employ. Choice of the ILP parameters is tackled here in an enumerative fashion: the ILP solver is run multiple times for each partition number bound in order to find which latency constraints yield the best solution, and it is repeated until an upper bound on the partitions number constraint is reached. A clear disadvantage of this way of operating, however, is that the ILP model is solved multiple times, thus increasing the computational burden.

In [145], a further modification of the approach is introduced: the operations inside each task are now allowed to be assigned to different partitions, thus refining the granularity at which the scheduling is performed. The authors stress here that, given the unbalance between the configuration and execution times, in order to obtain a good overall latency it is necessary to minimize the need for reconfiguration, and thus the number of temporal partitions identified.

The approach that they propose to this end is the *sharing* of the functional units instantiated in each temporal partition between different operations allocated to the same partition. The authors present as an example the application of their approach to the description of a matrix multiplier, showing that making use of sharing can yield a gain in execution latency. The approach proposed therein employs a resource-constrained force-directed list scheduling, the resource set of which is modified to signify the change in the chosen temporal partitioning in a local search-like fashion. A major issue that the authors are faced with is, once again, the remarkable unbalance between the time spent in computation and reconfiguration, the latter being orders of magnitude more relevant than the former, several seconds spent in reconfiguration vs. just microseconds needed for execution.

This observation leads the authors, in a further work [114], to explore the possibility of lessening the impact of reconfiguration time on total latency. They observe here that many data-intensive applications can be modeled as the subsequent application of some repeating tasks, in a loop-like fashion. If each one of the tasks were to be implemented as a temporal partition, then a reconfiguration would be necessary each time the application needed to begin execution of a new task. The possibility explored here is that, depending on the structure of the application, it might be possible to configure the device for each task type only once, by running the task on all of the data it needs to process before going on to the next configuration.

As an example, the JPEG compression algorithm is internally composed of four well-defined tasks (DCT, quantization, zig-zag, and Huffman encoding), each of which is executed independently on all of the image data to be compressed or on the output of the previous task. A problem foreseen by the authors with this approach is that, for some classes of algorithms, the intermediate data that the application would need to save to ensure proper communication between a task and the subsequent one could have a significant size. Thus, they propose two ways to cope with this issue: the first one is called *Final Data to Host*, in which each task is only run as many times as the on-board memory of the embedded system allows it to, after being swapped out for the next one, while the other is *Intermediate Data to Host*, which has each task run on all of the input data and generates all the outputs before being overwritten with the next temporal partition, even if costly data accesses to external memory are needed. Experimental data provided by the authors suggest, as expected, that the reconfiguration times can only be effectively hidden when the data intensiveness of the application is appropriately stressed, e.g., for the JPEG case only when the input image exceeds a certain size.

Other approaches, like the one proposed in [36, 35], deal with the possibility of using high-level languages as the means of writing system specifications. More specifically, the authors here focus on defining a way to implement a specification given as Java bytecode on a reconfigurable hardware platform. The specification's data flow graph is first traversed in order to produce an ordering of the nodes based on topological levels, with mobility of the nodes (`ASAP-ALAP`) used to break ties between nodes on the same level. From here on, the approach is pretty similar to the one presented in [150]. Unfortunately, little space is devoted to the discussion of reconfiguration time. Further work from the same authors [34, 38] tries to apply a concept that we have already exposed above for other approaches: the sharing between different operations of the functional units present in each temporal partition. The proposed way of proceeding here is to first create a temporal partition for each node on a critical path of the specification's data flow graph, and then try to fit the other nodes in a partition according to their dependencies and resource usage, taking advantage where feasible of resource sharing. In a subsequent step, an effort is made to try and merge bordering partitions.

In [33] the authors also propose a novel way to split a loop into several time partitions, which is useful when its body cannot fit onto the resources offered by the reconfigurable device. This, however requires a large number of reconfiguration steps, and thus greatly affects the system's latency, making it really only advantageous when there is no other way of fitting the specification onto the device.

A series of works related to the MorphoSys platform [128, 130] tries to approach dynamic reconfiguration with respect to the particular features of this architecture. The core of the architecture is made of an $8x8$ array of reconfigurable cells, each of which resembles a microprocessor's datapath and receives control information from a dedicated control register, which in turn loads its contents from the context memory. The task of controlling the behavior of the reconfigurable hardware is delegated to a dedicated RISC control processor. The proposed approach uses a library of *kernels* as its starting point, i.e., a collection of subprograms written in C each of which has its counterpart implemented as reconfigurable hardware. It is also assumed that a generic specification can be expressed as a loop on a certain sequence of kernels. Note that this assumption has been adopted in other works such as [114], presented earlier. The reason this approach is considered as only performing temporal partitioning is that each kernel is assumed to require the use of the whole processing array at once. Having said this, it appears clear that the key to optimizing the implementation is an appropriate way to manage the loops that the specification is composed of. Not unexpectedly, the first approach proposed is the modification of the loops' schedulings in order to execute all necessary runs of each kernel before reconfiguring the hardware to execute the next one, with the objective of minimizing the number of required reconfigurations. Different ways of rescheduling the kernels to this end are proposed in [130], taking into account the limitations imposed by the hardware architecture. Once the loops have been scheduled, the authors focus in turn on the generation of the control code, which states when context changes take place, and how data are transferred between subsequent partitions. A primary goal is the maximization of the overlapping between computation on part of the reconfigurable array and data transfer or context loading, along with optimal management of the memory in order to simplify the control code.

3.2.2 Spatial and Temporal Partitioning Approaches

In [195, 196, 194], Vasilko et al. deal with reconfigurable architectures that have, ideally, a reconfiguration time of less than one clock cycle. Their approach [195] is to sort the tasks composing the specification according to their `ASAP` and `ALAP` labels, and then using a normal list-based scheduler constrained in resource usage to the area available on the device. When the area is used up, functional units configured to implement tasks that are no longer needed are removed from the device, and replaced with new ones. The fragmentation that might arise from such usage of the device is unfortunately not dealt with. [194] is devoted to the presentation of a graphical tool to allow the designer to manually *pack* the functional units, both regarding their placement on the device and across partitions, and thus explore the design space.

An automatic approach is instead proposed in [196], where the authors try to explore the design space by taking advantage of a genetic algorithm, tackling task allocation to functional units, scheduling, and floorplanning all at once.

The authors of [144, 113, 145, 112, 114, 70, 78], cited above, also explore the possibility of extending their work to take advantage of partially reconfigurable architectures. Their main concern [70] is to hide the reconfiguration times: their approach is to divide the reconfigurable device into two partitions of fixed sizes. At a given time, one of the two parts is computing while the other one is being reconfigured and getting ready for its next task. It is assumed that the configuration of one of the two parts starts synchronously with the execution of the other one, which implies that in order to properly hide the reconfiguration times, the task assigned to a part has to be long enough in execution time to last at least as long as the configuration of the other part. In order to ensure this, the now well-known idea of embedding a loop inside each partition is proposed.

This is made harder, clearly, by the fact that each partition can only occupy half of the resources on the device, and thus can only fit smaller loop bodies.

The authors' focus changes in [78], where they try to minimize the computational effort needed in the design phase as opposed to optimizing the execution further, by exploiting a library of pre-placed and routed macros as functional units that the design phase thus only has to place on the device.

3.2.3 Regularity Extraction

Detecting regular patterns from the designer's specification is a promising approach to find an interesting solution to the partitioning problem for reconfigurable architecture. The detection of repeated structures, in fact, allows us to identify reconfigurable modules that are reusable, thus saving reconfiguration time. This extraction of regular structures from a given graph is actually a valuable step in different contexts, ranging from software engineering, e.g., in detecting duplicate code [120] both for plagiarism detection and for optimization purposes, to hardware areas such as VLSI design [121, 9, 153, 44, 159].

Extraction of regularity from a given specification is an approach holding great potential in the definition of a design flow for reconfigurable architectures: striving to *reuse* the configurable modules we identify is a promising way to lessen the impact of configuration times on the performance of the overall system. The following presents an overview of some approaches to regularity extraction presented in the literature.

Both [121] and [9], which work directly at the circuit level, propose *signature-based* approaches to describe the local structure of the design and thus recognize regular structures.

The authors of [9] try to recognize recurrent interconnection patterns between circuit elements from a known library, in order to optimize placement with respect to area and wire length. At the core of their proposed approach is the use of a hash function to summarize the features of each circuit element, taking into account the element's type and its interconnection pattern with its neighbors. The described approach aims at identifying blocks of regular functions among random instances that are not part of the datapath, in order to place them into the bit-slices to which they show the closest relation, thus saving communication costs.

The work presented in [121], also in the VLSI design field, has the goal of identifying functionally equivalent *slices*, i.e., subgraphs of the netlist generated by the system's HDL description, in order to ease the subsequent logic optimization step. The proposed approach starts with the identification of seed sets, i.e., initial groups of known equivalent gates, to be used as a starting point.

Seed sets are built considering heuristics closely related to the application field, such as considering successors of high-fanout nodes, or entities with similar names as generated by the HDL tool. At the core of the described algorithm is an expansion of the already known functionally equivalent slices driven by the comparison of the so-called *regularity signatures* between the corresponding entities in each of the slices. These signatures encode information about the structure that would be obtained by expanding the current slices. Basically, they are subgraphs that include the neighbors being considered for inclusion into the slices along with the edges connecting the new neighbors to entities already included in the slices. By comparing the signatures, the algorithm can ensure functional equivalence of the slices obtained after the expansion.

Another way to approach the regularity extraction problem is presented in [153], which also deals with integrated circuit design. Here, the focus of the work is on recognizing in the system graph instances of a given set of templates. The authors propose an approach that involves using a linear representation of the graphs, in the form of strings called *K-formulas* [22]. By representing the templates to be recognized as K-formulas, the authors formulate the matching problem as a string matching one, and propose ways to restrict the K-formula representation, which in itself would allow a given graph to have multiple legal string encodings, to obtain a unique representation for a given directed graph. The authors, however, assume that the template library is given and thus focus on the recognition of their instances in the circuit. The problem of generating templates is, on the other hand, given little space.

3.2.4 Tree-Shaped Recurrent Structure Detection

The work presented in [44] aims at detecting recurrent structures in circuit descriptions, with the purpose of aiding the subsequent hardware design phases. The input for the two algorithms proposed in the paper is an acyclic data flow graph. The two algorithms both deal with the same problem, i.e., recognizing isomorphic structures in a given data flow graph. In general, this has been shown to be an NP-complete problem, and the authors of [44] chose to deal with it by placing restrictions on the structure of the subgraphs to be identified. The two algorithms differ in the structure of said subgraphs, the first one dealing with *tree-shaped* ones, the second instead working with so-called

Single Principal Output Graphs (SPOGs) [44].

In the following, we give a detailed description of the tree-shaped subgraph approach, which was implemented as part of this work to allow early testing of the coherence of the data structures by employing an algorithm of proven validity. Let us define the concept of *template*, a recurrent structure repeated in the specification, and *instance*, as each of the repetitions of a given template in the specification.

The goal of the algorithm is the covering of the input graph by utilizing instances of tree-shaped templates, i.e., templates characterized by having a *single output*, and within which each of the nodes has a single output as well.

To describe the algorithm's features, let us first of all explain how the algorithm produces the templates to be used for covering the graph. This template generation step is described in Algorithm 1: therein we can see how the main loop goes through pairs of vertices, where the second always follows the first in topological order, and relies on the function $LargestTemplate$ to detect whether the two are roots of a common template.

The data structures involved in this algorithm are the following:

- G is the input data-flow graph;
- S is the output that the template generation produces, i.e., a set of templates;
- $template_{i,j}$ stores, for each two vertices i and j, the common template rooted at the two vertices, if any;
- $rootNodes_{template_i}$ is the set of nodes at which the instances of $template_i$ are stored.

The rationale behind the way this algorithm works is that, at any iteration, we are trying to see whether a common template exists that has i and j as its *roots*, i.e., the nodes that generate the *only output* of the tree. It is important to note how the instance of a template is fully identified by its root node: this simplification is allowed by the restriction we are enforcing on the shape of the templates we consider.

A point worth noting here is that the graph is traversed according to the topological ordering of its vertices. This means, at any given iteration of the loop the ancestors of the current nodes, if any, *will already have been visited.* This is a fundamental point, as we will see, on which the $LargestTemplate$ function is based.

Once the $LargestTemplate$ function has been computed, if it has detected no common template rooted at the two current nodes then the pair is simply discarded and the next one is considered. If instead a common template has been detected, then some bookkeeping has to be performed: the $template_{i,j}$ variable is updated to account for the fact that the current template has two instances rooted at nodes i and j. This information will be used later on, when the algorithm considers children of the current nodes.

Algorithm 1 Identify Templates (G)

```
Require: G is a DAG
  S ← ∅
  topologically sort the nodes of G
  for each node i of G do
    for each node j of G following i in topological order do
      currentTemplate ← LargestTemplate(i, j)
      if template ≠ nil then
        template_{i,j} ← currentTemplate
        if template ∈ S then
          rootNodes_{currentTemplate} ← rootNodes_{currentTemplate} ∪ {i, j}
        else
          S ← S ∪ currentTemplate
          rootNodes_{currentTemplate} ← {i, j}
        end if
      end if
    end for
  end for
  return S
```

At this point, all that remains to be done for the current pair is to see whether the template they have in common is a known one or if, instead, it is the first time such a template is encountered. In either case, at the end of the computation for the current pair the template will have been inserted in set S, and the current pair of nodes will have been added to the $rootNodes_{template}$ set.

Let us now consider how the *LargestTemplate* function (Algorithm 2) carries out its task. In order to understand its workings, we first have to clarify how a template is treated in this approach to the problem. Thanks to the fact that we are identifying tree-shaped templates, we can efficiently store only the type of the root node, and keep track of which other templates are *children* of a given one.

A template is as such composed of two elements[‡]:

- *rootFn* is simply the kind of operation performed by the root of the template;
- *childrenTemplates* is a list of children templates, i.e., templates rooted at ancestors[§] of the root node of the current template.

[‡]In the pseudocode explaining the algorithm, this is represented with a C-like dot notation for the sake of readability.

[§]This might be somewhat confusing, but we must keep in mind here that the term *tree-shaped* is used in a somewhat stretched way: we are in fact identifying templates that are shaped like reversed trees, i.e., their root is a node that has no children inside the template.

It is of basic importance to note here that, since the graph is visited in topological order, then the templates which are candidates to be children of the one being built have, at any time, *already been generated.*

Other data structures accessed in this function are the same used in algorithm 1, with the exception of $\alpha(\cdot)$, which we had defined as the operation type associated with a given node. Moreover, $\mu(\cdot)$ is implicitly accessed when ordering the ancestors according to the index of the input they provide. Given the rather strong assumptions about the kind of templates being generated, and the restriction to a DAG of the input graph, the way *LargestTemplate* works is rather straightforward. First of all, if the two nodes currently being considered represent different operation types, then they surely do not have any common template rooted at them. Otherwise, they do have a common template: *rootFn* is then set to the common operation type of the two current nodes. Next, the ancestors of the current nodes are taken into consideration, in order to see whether they have any previously identified template in common. Once again, keep in mind that due to the topological ordering of the graph it can be assured that these ancestors have already been visited at this time. The ordering of the inputs also has to be taken into account here: the template being identified can be expanded to include two ancestors only if these ancestors provide the same input for the current nodes. Here is where the $template_{i,j}$ variables are actually used: since templates for the ancestors have surely already been detected, then it can be assured that, if two ancestors providing inputs at the same index share a template, it will be found in the respective $template_{i,j}$ variable. If such a template is found, then it is added to the *childrenTemplates* set for the template currently being generated.

An important point to note is that *LargestTemplate* only adds a child template *if the ancestors at which it is rooted have a single output.* This is done to ensure that the identified instances of a given template are nonoverlapping, and to restrict the algorithm to maintain the tree structure when generating templates.

Once all the ancestors have been considered, the current template is ready and can be returned.

As an example of how this procedure for template identification works, consider Figure 3.3. Note that the figure is intentionally incomplete, i.e., it does not show *all* of the generated templates, for the sake of clarity. The templates are numbered in the same order as they would be generated, i.e., ancestors first. Once it has generated templates 1 and 2, the algorithm would then generate template 3 and, noticing that ancestors of the root nodes of template 3 are roots of the same previously identified templates, would add them to the *childrenTemplates* for template 3.

This concludes the explanation of how this algorithm identifies templates in the graph. However, how the algorithm performs the following step, i.e., the covering of the graph using the generated templates, still has to be exposed. Algorithm 3 shows the approach chosen by the authors in [44] for this task. First of all, template identification is carried out as outlined in algorithm

Algorithm 2 LargestTemplate (u, v)

if $\alpha(u) \neq \alpha(v)$ **then**
  return *nil*
else
  $sharedTemplate.rootFn \leftarrow \alpha(u)$
  for each pair $(a_{u,k}, a_{v,k})$ of ancestors of u and v providing inputs at the same index k **do**
    if $(a_{u,k}, a_{v,k}$ only have one child$) \wedge template_{a_{u,k},a_{v,k}} \neq nil$ **then**
      add $template_{a_{u,k},a_{v,k}}$ to $sharedTemplate.childrenTemplates$
    end if
  end for
  return $sharedTemplate$
end if

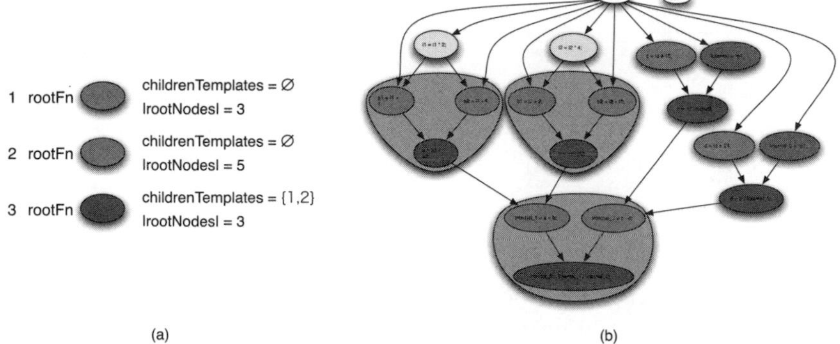

Figure 3.3: Example of how tree-shaped templates are generated. (a) shows the way the generated templates are stored, while (b) shows where the identified templates have their instances in the graph.

1. Then, one of the identified templates is chosen according to a heuristic, represented here as a *ChooseTemplate* function, which represents the *fitness* a template has to being implemented as one of the configurable units. Examples of heuristics used for template choice include, choosing the largest templates (in terms of hardware area used) first - the so-called *Largest Fit First (LFF)* heuristic, or privileging the templates that recur more often, in the *Most Frequent Fit (MFF)* heuristic. The winning template according to the chosen heuristic is extracted, and its instances removed from the graph. The choice is then repeated on the remaining templates in the part of the graph still to be covered, until there are no more available templates to choose from.

Algorithm 3 Graph Covering Using Tree Templates

repeat
  $templates \leftarrow IdentifyTemplates\,(G)$
  $ChosenTemplate \leftarrow ChooseTemplate\,(templates)$
  $CoveredVertices \leftarrow \emptyset$
  for each node n in $rootNodes_{ChosenTemplate}$ **do**
    if $vertexSet\,(ChosenTemplate, n) \cap CoveredVertices = \emptyset$ **then**
      create a new cluster containing $vertexSet\,(ChosenTemplate, n)$ in the partitioned graph
      add $vertexSet\,(ChosenTemplate, n)$ to $CoveredVertices$
    end if
  end for
  remove the nodes in $CoveredVertices$ from G
until $templates = \emptyset$

3.2.5 The DRESD Partitioning-Based Approach

Looking at the application domain we see that reconfiguration time is still much larger than execution time: about two orders of magnitude, even if the distance is getting ever shorter as technology advances. One is therefore prompted not to *hide* reconfiguration, but rather to *reduce* it. Doing this in a substantial way will have an impact on total latency far more decisive than any fine-tuning of the scheduling. In order to achieve this reduction in reconfiguration we will try to create $\mathcal{S}$ in such a way that reuse is possible: if many sets S in $\mathcal{S}$ are mapped onto the same EU, we can execute them on the same RUs with a single starting reconfiguration. Finding such *equivalent* sets amounts to finding isomorphic distinct subgraphs in the specification DFG $\langle O, P\rangle$.

3.2.5.1 Core Generation

The problem of finding the two largest non-overlapping isomorphic subgraphs of a given graph, ISOMORPHIC SUBGRAPHS, has not been studied as extensively as related problems, like GRAPH ISOMORPHISM or COMMON SUBGRAPHS (see [220, 222]), and it is not reducible to them. The only papers dealing with it are [12, 13, 14]: here the authors solve the problem in linear time for the case of trees (which is not useful to us) and give a heuristic for the general case of a non-colored, non-directed graph. In this section, a set of nodes $S \in \mathcal{S}$ will be also used to denote the unique *node-induced subgraph* of $\langle \mathcal{S}, \mathcal{P} \rangle$ induced by the nodes of S. Also, a neighborhood of S will be any connected subgraph of $S \in \mathcal{S}$ containing S.

The algorithm proceeds as follows:

1. Build a collection of *small* pairs of disjoint isomorphic subgraphs $\mathcal{I}$;
2. Choose a pair $(S^a, S^b) \in \mathcal{I}$ and remove it from $\mathcal{I}$
   (a) build their non-overlapping neighborhoods $\mathcal{N}(S^a)$ and $\mathcal{N}(S^b)$. If this is not possible, goto 3;
   (b) perform a maximum bipartite matching between $\mathcal{N}(S^a)$ and $\mathcal{N}(S^b)$, where the weight of an arc is computed as a "similarity measure" of the two nodes it connects;
   (c) for all the matched pairs, check if adding their respective nodes to S^a and S^b really preserves the isomorphism. If so, add them to S^a and S^b. Goto 2(a)
3. Save the maximal isomorphic non-overlapping subgraphs S^a and S^b. Goto 2.

The algorithm has two major drawbacks. The initialization phase requires several initial pairs of isomorphic subgraphs to be used. Moreover, ensuring that the initial sets are isomorphic normally amounts to choosing singletons, which means that the enlargement process will run for a long time. Also, the computation of good similarity weights is time consuming, since it uses graph-theoretical measures such as distance and so on. We are in a position, however, to improve on the situation using the peculiarities of the specific domain.

Our graphs exhibit more structure than that considered in [12, 13, 14]. In fact they are:

- *colored* (the nodes correspond to different actions to perform. More specifically, to each operation o an action $a_o \in A$ is associated);
- *oriented* (the arcs correspond to a directed data flow);
- the ingoing arcs might be *ordered* (the action performed by each node on the incoming data can be non-commutative).

These facts yield some optimizations to the steps of the algorithm.

1. Our approach to the determination of the initial sets is as follows: suppose we consider an arc $e = o_1 \rightarrow o_2$ that serves as n–th input for o_2, where operation o_1 executes action a_1 while o_2 executes action a_2. We create a set $I^n_{a_1 a_2}$ and store e in it. Iterating this for all the arcs takes a linear time in the number of the arcs, and results in $|A|^2 N$ sets (where $|A|$ is the number of possible actions and N is the maximum number of inputs to a node) made up of isomorphic subgraphs of $\mathcal{S}$. All the elements of these sets are pairwise suitable to be used as initial points for the algorithm.

2(a,b). Algorithm 4 summarizes the behavior of the expansion procedure starting from two arcs identifying isomorphic subgraphs of 2 vertices each. As for the matching, we build an auxiliary bipartite graph between S^a and S^b, introducing an edge between any two compatible nodes, that is, those with same fan-in, fan-out, action and input-order.

2(c). The local isomorphism check performed to decide whether to add a new pair of nodes to the isomorphism bijection has to take into account that we are working with *node-induced* subgraphs, i.e., subgraphs defined by a given set of nodes, and including all the arcs from the parent graph having both source and destination in the subgraph. It is therefore carried out by building two lists, containing the arcs added to each of the subgraphs being expanded by the addition of the new node pair, and comparing them to ensure they indeed satisfy the definition of isomorphism.

An example of execution of the algorithm is shown in Figure 3.4.

3.2.5.1.1 Local Isomorphism Check Let us now consider the local isomorphism check step, which was mentioned above. At first glance, it may seem that making sure the operation types for two neighbors match is enough to ensure the isomorphism of the clusters with the neighbors added to them. This is not quite true: we must keep in mind, in fact, that we are considering *node-induced subgraphs*, i.e., subgraphs that include the vertices in their vertex set and *all the edges from the original graph that have both source and destination in the subgraph.* Figure 3.5 illustrates this problem with a simple example. In the figure, the six nodes surrounded by the grey area are already part of the two isomorphic clusters, which consist of three nodes each. The nodes labeled c_1 and c_2, as in the algorithm's pseudocode, are part of the pair that has been dequeued from P and whose neighbors are being considered for addition to the clusters. After the matching problem has been solved, the nodes labeled m_1 and m_2 have been matched. Their operation type is the same, but adding them to the clusters would *not* produce isomorphic clusters of size 4!

Notice, in fact, how adding m_1 and m_2 to their respective subgraphs would also cause the addition of the edges from m_1 to the light blue node and from m_2 to the yellow node. The resulting subgraphs would then not be isomorphic.

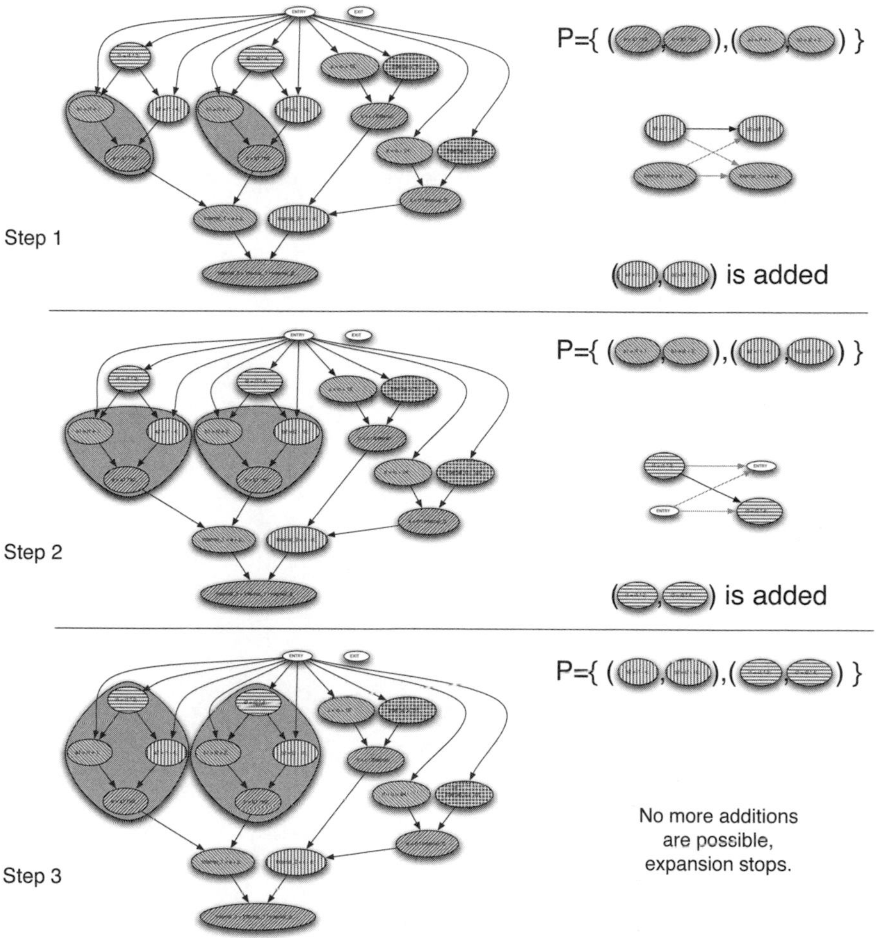

Figure 3.4: A sample run of the candidate generation.

Algorithm 4 ExpandTemplate $((s_1, d_1), (s_2, d_2))$

$S^a \leftarrow \{s_1, d_1\}$
$S^b \leftarrow \{s_2, d_2\}$
empty queue P
enqueue $(s_1, s_2), (d_1, d_2)$ into P
while P is not empty **do**
  $(c_1, c_2) \leftarrow$ dequeue from P
  $\mathcal{N}^a \leftarrow EnumerateNeighbors(c_1)$
  $\mathcal{N}^b \leftarrow EnumerateNeighbors(c_2)$
  formulate the matching problem between the elements of $\mathcal{N}^a$ and $\mathcal{N}^b$
  for each pair (n_1, n_2) in $\mathcal{N}^a \times \mathcal{N}^b$ **do**
    $w_{n_1,n_2} \leftarrow ComputeMatchingWeight(n_1, n_2)$
  end for
  $matching \leftarrow SolveMatching()$
  for each match (m_1, m_2) in $matching$ **do**
    if $IsomorphismCheck\left((S^a \cup \{m_1\}), \left(S^b \cup \{m_2\}\right)\right)$ **then**
      $S^a \leftarrow S^a \cup \{m_1\}$
      $S^b \leftarrow S^b \cup \{m_2\}$
      enqueue (m_1, m_2) into P
    end if
  end for
end while

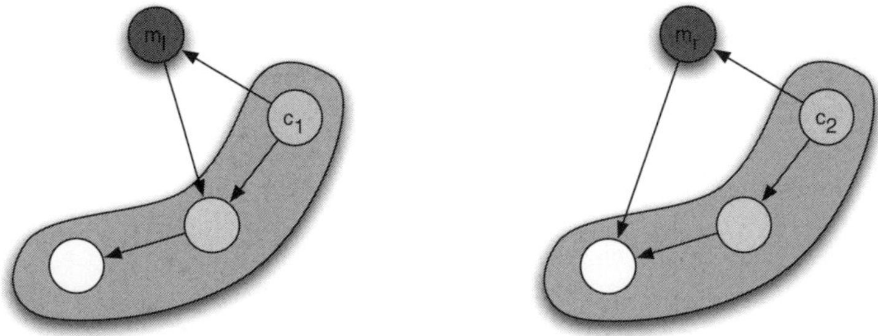

Figure 3.5: Local isomorphism check: example.

In order to avoid this kind of issue, the local isomorphism check has been implemented as shown in Algorithm 5. The idea behind the check that is performed is indeed pretty simple: we build two sets, E_1 and E_2, which represent the edges that would be added in the left and right clusters, respectively (*left* and *right* are intended as the left and right sets of the bipartite matching), if we added the node currently being checked. Each edge is represented by a triple, $\langle inout, pair, \mu\rangle$, in which:

- *inout* can take two values: *in* if the edge being described has the newly added node as its destination, or *out* if the opposite happens
- *pair* is the pair of nodes in the isomorphism bijection to which the node already in the cluster which is the source (or sink) of the current edge belongs
- μ is the input number provided by the edge

The algorithm for local isomorphism check simply populates the two sets, with edges to and from the nodes being considered for addition to the clusters. When the sets have been populated, a simple comparison is performed: if the contents of the two sets are the same, then the local isomorphic check is passed. Otherwise, addition of the two new nodes is rejected.

Algorithm 5 IsomorphismCheck

$e_l \leftarrow \emptyset$
$e_r \leftarrow \emptyset$
for each edge $(m_l \rightarrow o) \in edges(G)$, with $o \in V_1$ **do**
  $e_l \leftarrow e_l \cup \langle out, (o, V(o)), \mu(m_l \rightarrow o)\rangle$
end for
for each edge $(m_r \rightarrow o) \in edges(G)$, with $o \in V_2$ **do**
  $e_r \leftarrow e_r \cup \left\langle out, \left(V^{-1}(o), o\right), \mu(m_r \rightarrow o)\right\rangle$
end for
for each edge $(o \rightarrow m_l) \in edges(G)$, with $o \in V_1$ **do**
  $e_l \leftarrow e_l \cup \langle in, (o, V(o)), \mu(o \rightarrow m_l)\rangle$
end for
for each edge $(o \rightarrow m_r) \in edges(G)$, with $o \in V_2$ **do**
  $e_r \leftarrow e_r \cup \left\langle in, \left(V^{-1}(o), o\right), \mu(o \rightarrow m_r)\right\rangle$
end for
if $e_l = e_r$ **then**
  return *true*
else
  return *false*
end if

3.2.5.2 Evaluation Functions

Once the core generation phase is over, the generated cores have to be evaluated to provide the following phases with a means of deciding which ones to choose for implementing the system: the scheduling and allocation step will decide whether to implement the cores as hardware modules or run them as hardware tasks. The whole graph representing the specification will have to be covered by cores, regardless of whether they will be implemented as hardware or software; however, it is not until the cover generation phase that such a condition will be met. This stage offers some metrics relevant for this decision:

- Core size estimates, either in graph theoretical terms (i.e., number of nodes) or more accurately in terms of forecasted device area usage[¶], useful to drive the following choices especially in conjunction with knowledge about the resources offered by the specific device considered. The current approach to device area usage estimation assigns a CLB (*Configurable Logic Block*) weight to each operation type, and estimates communication overhead by means of a multiplicative factor;

- Configuration time estimates, of obvious importance to our objective since one of the primary aims of this research is to try and reduce the impact of reconfiguration times. These estimates are in close relation with the area ones since there often exists, at least with FPGA technologies, a linear relation between occupied area and configuration time;

- Latency estimates, i.e., an estimate of the time a given core will take to execute its functionality once configured on the target device. This is of course essential to the subsequent scheduling phases, but can nonetheless be useful during the core choice phases, e.g., to choose between two cores with identical configuration time but different latency: which one will yield the most advantage if chosen?

Other ways to characterize the identified cores were defined, among which the so-called *Hierarchical Template Graph*, HTG in short, is worth noting. The result of several runs of an algorithm solving Isomorphic Subgraphs is a set $\mathcal{V}$ of pairs of isomorphic subgraphs. This set, though, is highly unstructured: it might contain, for instance, two pairs whose elements are isomorphic (say $\{S^a, S^b\}, \{S^a, S^c\} \in \mathcal{V}$), which would be better represented as a *single* set $\{S^a, S^b, S^c\} \in \mathcal{V}$. Also, notice that even if two sets S^a, S^b in some pairs of $\mathcal{V}$ are not isomorphic, it might well be that there exist $S^{a'} \subset S^a$ and $S^{b'} \subset S^b$ that indeed *are* isomorphic. For instance, say we have $\mathcal{V} = \{\{S^a, S^b\}, \{S^c, S^d\}\}$ and that there exists $S^{a'} \subset S^a$ isomorphic to

[¶]An accurate measure of device resource usage can only be obtained further down the design flow, due to the nondeterminism of some of the lower level phases.

S^c. For scheduling purposes we are not sure whether we prefer to use less isomorphic subgraphs of greater size ($\{S^a, S^b\}$) or more isomorphic subgraphs of smaller size ($\{S^{a'}, S^{b'}, S^c, S^d\}$, where $S^{b'} \subset S^b$ isomorphic to $S^{a'}$ must exist because S^a is isomorphic to S^b).

For this reason we build a *Hierarchical Template Graph*, HTG, whose nodes contain a collection of isomorphic subgraphs. Whenever two subgraphs are isomorphic they share the same *structure*, so we introduce the term *template* to identify the common structure of those subgraphs. Moreover, every arc starting from a set S and going to a set S' implies that the template for S (and hence any graph in S) is a subgraph of the template for S' (and hence of any graph in S').

If we now want to know how many copies of a subgraph S there are in the original graph $\mathcal{S}$ we can look at the representative node for S in $\mathcal{V}$ and traverse all the *children* of that node, since they are bound to contain graphs $\bar{S}$ that have subgraphs isomorphic to the representative and hence to S. Therefore we can say that in the HTG two templates have a common ancestor if they *contain* it. This situation is especially true of the core generation approach we exposed above, which dealt with the progressive expansion of an initial *seed* graph and kept track of each of the intermediate results, but it definitely can occur in other approaches as well. The purpose of the HTG structure is to explicitly represent such relations, in order to allow us to take advantage of the structural similarity of the cores in their choice. Figure 3.6 shows an example of the structure offered by the HTG, in which it is shown how the *seed* templates of size 2 are parts of larger templates of size 3, which in turn are subgraphs of the size 4 template. The templates used are taken from the example used earlier for the core generation phase.

The HTG can improve the reconfiguration process in two different ways. On one hand it allows the scheduling algorithm to compute a fairly precise trade-off between the size of the templates to use on the FPGA and the number of times these replicas can be used.

On the other hand, the scheduling algorithm can benefit from knowing that a certain template can be configured on the FPGA as a *modification* of another template rather than by blindly reconfiguring the CLBs. As an example, consider two nodes, n_1 and n_2, $n_1, n_2 \in$ HTG, that have a common ancestor, $a_i \in$ HTG. This means that n_1 and n_2 contain two isomorphic subgraphs. It is then possible to specify the reconfiguration process as the *difference* between n_1 and n_2, thus producing a smaller reconfiguration bitstream to download on the FPGA.

As a proof-of-concept, consider Table 3.1, which shows some reconfiguration time results computed on a Xilinx xc2vp7 FPGA. It is easy to see that the reconfiguration time grows linearly with the size, in number of frames, of the bitstream. Therefore, the ability to maximize the resource reuse is going to drastically decrease the time needed to partially reconfigure the device.

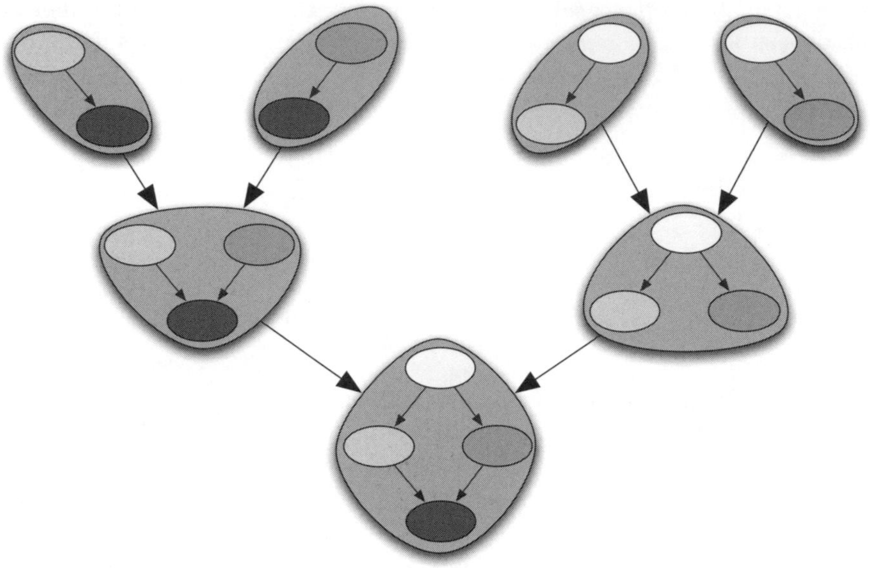

Figure 3.6: Hierarchical Template Graph (HTG): example.

Table 3.1: Dimension vs. Reconfiguration Time

Label	Reconfiguration frames	Configuration time (ms)
RC1	60	32.700
RC2	68	34.203
RC3	132	60.045
RC4	148	64.140
RC5	182	76.049

RC1 to RC5, shown in Table 3.1, represent the bitstreams used to implement a given specification on a reconfigurable device. Suppose we have an isomorphic relation, computed using the HTG, between some of them, i.e., RC4 and RC5 have as common ancestor RC3. Suppose also that a data dependency exists between RC3 and RC4 and between RC4 and RC5, so that RC5 has to wait for RC4 and RC4 for RC3. In a scenario with just one feasible area to configure all the RCs there is no way to hide the reconfiguration time needed to switch between the corresponding three configurations. Therefore the configuration time is computed as: 60.045ms+64.140ms+76.049ms = 200.229ms.

Consider now the case where the information provided by the HTG has been taken into account to compute the reconfiguration bitstream. In this scenario the reconfiguration time needed to configure RC3 is still 60.045ms, but the time required to switch between RC3 and RC4 is no longer 64.140ms but something close to 4ms, and the one to switch RC4 to RC5 is closer to 20ms, therefore the overall configuration is now 60.045ms + 4ms + 20ms $\approx$ 84ms.

All this information is already available thanks to the HTG. At present it is not as yet used in the scheduling process, but its exploitation is underway.

Building this tree may seem complicated, but in practice it is sufficient to identify every node of the HTG with a representative list of operations from the original graph. We start with an empty tree and add in turn the pairs in decreasing size order. If in the graph there is a node with the same list of operations (we can order the list of operation in some way to speed up the comparison), we add the pair there. If the list of operations is a subset of the list in an existing node, we create a new child node.

3.2.5.3 Core Choice

The core choice phase takes as its input the set of cores produced during the core generation phase, along with the information provided by the evaluation functions mentioned above, and chooses which of the proposed cores will be used in the reconfigurable implementation of the designer's system.

Currently, we have implemented a *greedy* approach: we choose the core maximising one of the fitness functions provided by the previous core evaluation phase, add it to the list of cores composing our reconfigurable implementation, then choose the next ones until no more usable cores exist. We have to point out that we are looking here for solutions in which the chosen cores *do not overlap* with one other. We enforce this constraint by eliminating from the list of usable cores, after each choice, the ones that overlap with the nodes already contained in the core we just chose.

A specialization of the core choice algorithm taking into consideration the organization in templates and instances of the cores, which is produced by our partitioning policy, is shown in Algorithm 6. Since, after the choice of which template to use, we remove instances from other templates which overlap with the one we just chose, we also get rid of the templates that have become useless in that they have no more instances left, or only have one (at which point, they do not expose any regularity anymore and as such can be discarded).

The selection of which core to use, represented by the *selectTemplate*() call in Algorithm 6, can be carried out according to different strategies, among which we can cite:

- Consider the core that has the largest size first, also known as *LFF (Largest Fit First) heuristic* [44]. The point of such choice is to allow greater optimization possibilities to later stages of the design flow, i.e., intra-module placement and routing, but has the potential adverse effect of choosing templates so large that they could use up a big part of the

Algorithm 6 Core Choice Step

```
while there still are usable instances in template_i for at least one i do
  k ← selectTemplate()
  generate a task graph node for each of the usable instances in template_k
  for each i do
    for each instance u ∈ template_i do
      if u ∩ u_k ≠ ∅ for any u_k ∈ template_k then
        remove u from template_i
      end if
    end for
    if |template_i| < 2 then
      delete template_i
    end if
  end for
end while
```

device, thus leaving little room for configuring more modules at the same time.

- Consider cores with the most instances first (clearly, this heuristic is especially fit to work with partitioning policies that expose recurrent structures), also known as the *MFF (Most frequent Fit First) heuristic* [44]. Conversely, this approach allows for more moderate area usage in that it tends, in principle, to maximise reuse of the identified units possibly at the expense of a worse schedule length in the implemented system.

In addition to these rather basic policies, however, we can define some that are more specifically tailored to our problem, i.e., the final goal of implementing our reconfigurable system on a physical device such as a FPGA with its constraints related to the minimum size of a reconfigurable unit, the total area availability and the reduced communication resources.

As an example, we might evaluate actual area usage on the device by using the following formula:

$$A_{used} = \left\lceil \frac{A_{core}}{Size_{rec}} \right\rceil Size_{rec}$$

i.e., the actual area used on the device by a given core is the area effectively occupied by the core's logic rounded up to the next reconfigurable unit.

3.2.5.4 Instance Choice

A noteworthy point exposed in Algorithm 6 is the concept of *usable* instances of a template, which arises from our choice of limiting ourselves, for the time being, to considering non-overlapping covers of the graph. While it would be

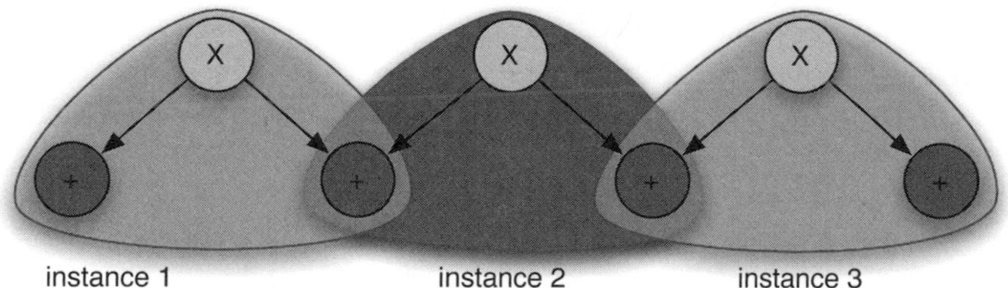

Figure 3.7: Instance choice: example.

desirable for us that all the instances within a template be non-overlapping, it is unfortunately not the case, despite us preventing the pair of instances from overlapping in each run of the template expansion. This, in fact, does not ensure that the instances generated *in different runs* of the expansion, which only works on a pair at a time, do not overlap. Consider as an example the case shown in Figure 3.7: in such a situation, we should be careful about utilizing a greedy approach since choosing instance 2 first would make us discard both instances 1 and 3. A more promising solution, here, is to choose instances 1 and 3, discarding instance 2.

In order to make the algorithm behave as expected in these situations, and thus try to select the maximum number possible of instances of the same template, an ILP problem is formulated for the choice of the usable instances of each template, representing the *conflicts*, i.e., overlaps between the different instances, which if overlapping make their choice mutually exclusive in our approach.

The formulation for the example in Figure 3.7 follows:

$$\max x_1 + x_2 + x_3$$
$$\text{subject to} \begin{cases} x_1, x_2, x_3 \in \{0, 1\} \\ x_1 + x_2 \leq 1 \\ x_2 + x_3 \leq 1 \end{cases}$$

which is a binary problem, where x_i is 1 if the ith instance is chosen, 0 otherwise. The formulation is rather straightforward, with the objective function simply stating that we desire the solution to use as many instances as possible, and the constraints formalizing the mutual exclusions dictated by the overlaps between the instances.

3.2.5.5 Covering Set Identification

The next phase is the covering set identification step which deals, as briefly pointed out before, with the fact that the chosen cores might not contain all

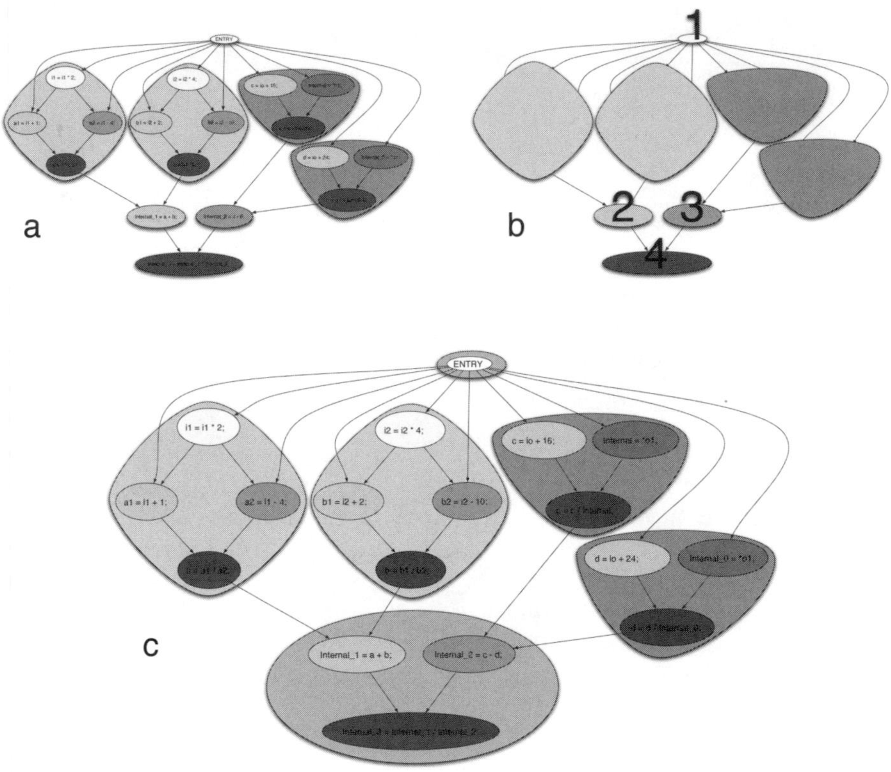

Figure 3.8: Covering Set Identification: (a) status after core choice, (b) order in which the covering set identification considers the uncovered nodes, (c) final status, partitions have been created for the previously uncovered parts.

the operations contained in the original specification. Basically, two ways exist to overcome this obstacle, which involve either *enlarging* some of the cores to include the uncovered operations or generating *new cores* which contain these operations. Given our policy of detecting repeated structures, we chose to act according to the latter strategy, since modifying the detected cores would mean losing the isomorphism between their instances, ultimately wasting the information produced by our partitioning phase.

Let us refer to Figure 3.8 to explain how we carry out this task, with reference to the example of Figure 3.4, where different instructions have been represented using different filler colors:

- Figure 3.8(a) represents the status of our example specification after the core choice phase, having applied our isomorphism detecting partitioner and an LFF heuristic for the core choice‖, with a simple size measurement based on the number of nodes contained in each core. Note how parts of the specification are uncovered by the chosen cores, i.e., the *entry* node at the top and the three nodes at the bottom;

- Figure 3.8(b) shows how the covering set identification algorithm traverses the graph to generate the new subsets: the graph is traversed in breadth-first order, with the aim of exploring the remaining nodes *by level* (nodes which have already been covered by the previous steps are simply skipped). The algorithm then adds the nodes to the current subset as they are visited, with the following exceptions:
  - to avoid introducing cycles in the output *collapsed* description, we enforce that any new subset may *not* have both incoming arcs *from* a chosen core *and* arcs *to* a chosen core. If adding the next node would violate this condition, a new subset is generated;
  - if the current subset reaches a size limit configurable by the user, a new one is created.
- Figure 3.8(c) represents the final status of the partitioned specification, with the newly created subsets shaded in gray.

3.3 Scheduling Techniques

We can now proceed with the scheduling phase.

Section 3.3.1 presents a short overview of the state of the art, presenting an ILP formulation proposed in [16] and [17], which has been used as the starting point for our formulation.

Then we describe three different approaches. The first one, in Section 3.3.2, is an exact ILP formulation, while the other two, in Sections 3.3.3.1 and 3.3.3.2, are heuristic methods.

3.3.1 Related Work

The task scheduling problem, working with a partial dynamic reconfigurable architecture, is similar to the one proposed in [16] and [17]. In [16] and [17] the authors consider the problem of partitioning and scheduling a task dependence

‖Largest Fit First, which chooses cores simply in decreasing size order. Other heuristics are MFF (Most frequent Fit First), which chooses the core that has most instances; and the two versions of CWM (Communication Width Metric), which try to minimize communication with the other cores or to maximize the communication inside the core.

graph onto an architecture defined using a general purpose processor (GPP) and reconfigurable devices where tasks can be executed both as software onto the GPP or as hardware cores using the reconfigurable resources. They propose a heuristic HW/SW partitioning and scheduling algorithm based on the well known KLFM heuristic. The limitation of these works is that they do not consider all the features available to partially dynamically reconfigurable devices, i.e., *module reuse* is rarely considered, and *anti-fragmentation techniques* are almost always ignored. In [105] the authors present an ILP formulation considering a task graph, partitioned into time–partitions, defined only with few task types achieving results similar to [113].

One of the first attempts to take into consideration partial reconfiguration combined with *configuration prefetching* is presented in [15]. The applications that can be scheduled with this approach are linear multi-task applications: they are a linear sequence of tasks and each task takes as input the results of the previous task; this kind of applications normally deals with large amounts of data, e.g., image processing. The goal in [15] is to define a specific methodology for scheduling the tasks of these applications in order to reduce the overall completion time. The same authors present in [18] an enhanced solution for the same problem: PANGRAN tries to reduce the total execution time using two different techniques. One, called *simple fragmentation reduction*, places the new task in the first available area on the FPGA in the opposite side of the FPGA with respect to the location of the previous task. The other, called *exploiting slack in reconfiguration controller*, is a local optimization that postpones the reconfiguration of a task in a position selected by the previous technique if this reconfiguration causes a delay in the subsequent task execution time. Furthermore, a task is replicated if and only if the replication gives an improvement in the execution time of the subsequent task considered as unreplicable; this technique is called *static pruning*. For replicated tasks there is a specific placement policy called *dynamic granularity selection*: placing multiple copies of a task in adjacent positions can reduce the fragmentation of the area of the FPGA; moreover, if it leads to a temporal improvement, the first copies of a task are stopped early in order to overlap the reconfiguration of the following task with the final execution of the last copies of the considered one. The authors do not consider the memory management in a context of parallel tasks that work on different portions of the same data; moreover, the memory available for the tasks is not quantified and this could be a problem.

In [71] a reconfiguration sequence manager is discussed, using as target architecture a partially dynamically reconfigurable system (partially dynamically reconfigurable FPGA systems or multi-FPGA systems). The proposed algorithm receives as input an already scheduled task graph and it aims at scheduling all these tasks onto the target architecture in order to minimize only the total reconfiguration time needed. The assumptions are the following: (a) every task has a fixed size equal for each task, with the same reconfiguration time; (b) on the target architecture there are K fixed positions where to put a task.

Under these assumptions the algorithm finds a provable optimal solution [71], and the techniques used are based on the *off-line paging* solved by *Least Imminently Used (LIU)* algorithm. This approach does not take into consideration resource reuse and different task sizes. Furthermore, only the total reconfiguration time is minimized, without considering the execution time and the reconfiguration time impact described, for example, in [63] for a similar architecture.

3.3.2 ILP Formulation

The previous model, though useful for analysis purposes, is too complicated to be solved with an ILP solver.

It is imperative, thus, to simplify it. This is done mainly by getting rid of the degree of freedom offered by partitioning and mapping, that is, we choose the specific partition $\mathcal{S}$ and for each subset $i \in \mathcal{S}$ (*task*) we also choose the EU to implement it.

Moreover, we consider 1D reconfiguration: the set U of RUs is the set of columns of the FPGA . Therefore, all RUs have the same ρ_u (conventionally, $\rho_u = 1$). Each task must be assigned to a sequence of adjacent columns and two reconfigurations cannot take place simultaneously.

We are then brought to consider the following simplified model, in which operator $[\cdot]$ takes value 1 if $[\cdot]$ its argument is true and 0 otherwise.

3.3.2.1 Constants

- a_{ij} := [tasks i and $j \in \mathcal{S}$ are mapped onto the same EU] (by convention, $a_{ii} = 1$);
- l_i := latency of task $i \in \mathcal{S}$;
- d_i := time needed to reconfigure task $i \in \mathcal{S}$;
- r_i := number of RUs required by task $i \in \mathcal{S}$,

Notice that index i replaces index e because the mapping of tasks on EUs is univocal.

The scheduling time horizon $T = \{0, \ldots, |T| - 1\}$ is large enough to reconfigure and execute all tasks. A good estimate of $|T|$ is obtained via a heuristic.

3.3.2.2 Variables

- p_{ihk} := [task i is present on the FPGA at time h and column k is its leftmost one];
- $\bar{t}_{ih}$:= [the reconfiguration of task i starts at time h];

- m_i := [task i exploits module reuse];
- S_i^{on} := arrival time of task i on the FPGA;
- S_i^{off} := last time instant when task i is on the FPGA;
- t_{e} := overall execution time.

3.3.2.3 Constraints

The constraints marked with an asterisk are written with the if-then transformation (see [210]).

NO OVERLAP CONSTRAINTS A column can be used by a single task at a time:

$$\sum_{i \in \mathcal{S}} \sum_{l=\max(k-r_i+1,1)}^{k} p_{ihl} \leq 1 \quad h \in T, k \in U \tag{3.2}$$

SINGLE COLUMN CONSTRAINTS A task cannot be present on the FPGA with different leftmost columns:

$$p_{ihk} + p_{iml} \leq 1 \quad i \in \mathcal{S},\ h, m \in T : k, l \in U : k \neq l$$

This constraint can be strengthened as follows:*

$$p_{ihk} + \sum_{l \in U \setminus \{k\}} p_{iml} \leq 1 \quad i \in \mathcal{S},\ h, m \in T, k \in U \tag{3.3}$$

SPACE-ON-THE-RIGHT CONSTRAINTS The leftmost column of task i cannot be one of the last $r_i - 1$ columns:

$$p_{ihk} = 0 \quad i \in \mathcal{S}, h \in T, k \geq |U| - r_i + 2 \tag{3.4}$$

ARRIVAL TIME CONSTRAINTS The arrival time must not exceed the first time step for which p is 1:*

$$S_i^{\text{on}} \leq h \sum_{k \in U} p_{ihk} + |T| \left(1 - \sum_{k \in U} p_{ihk}\right) \quad i \in \mathcal{S}, h \in T \tag{3.5}$$

LEAVING TIME CONSTRAINTS The leaving time must not precede the last instant for which p is 1:

$$S_i^{\text{off}} \geq h \sum_{k \in U} p_{ihk} \quad i \in \mathcal{S}, h \in T \tag{3.6}$$

NO-PREEMPTION CONSTRAINTS A task is present on the FPGA in all time steps between the arrival and leaving time:

$$\sum_{h \in T} \sum_{k \in U} p_{ihk} = S_i^{\text{off}} - S_i^{\text{on}} + 1 \quad i \in \mathcal{S} \tag{3.7}$$

Precedence constraints Precedences must be respected:

$$S_j^{\text{off}} - l_j \geq S_i^{\text{off}} \quad (i,j) \in \mathcal{P} \tag{3.8}$$

Task length constraints A task must be present on the FPGA at least for its execution time plus (if no module reuse occurs) its reconfiguration time (reconfiguration prefetching is allowed, that is, the execution is not bound to follow immediately the reconfiguration):*

$$\sum_{h \in T} \sum_{k \in U} p_{ihk} \geq l_i + (1 - m_i) d_i \quad i \in \mathcal{S} \tag{3.9}$$

Reconfiguration start constraints Each task has a single reconfiguration start time or none (if it exploits module reuse):

$$\sum_{h \in T} \bar{t}_{ih} = 1 - m_i \quad i \in \mathcal{S} \tag{3.10}$$

Reconfiguration starts as soon as the task is on the FPGA:*

$$-|T| m_i \leq S_i^{\text{on}} - \sum_{h \in T} h \bar{t}_{ih} \leq |T| m_i \quad i \in \mathcal{S} \tag{3.11}$$

Reconfiguration overlap constraints Two reconfigurations cannot take place simultaneously:

$$\sum_{i \in \mathcal{S}} \sum_{m=\max(1, h-d_i+1)}^{h} \bar{t}_{im} \leq 1 \quad h \in T \tag{3.12}$$

Starting time constraints The starting instant is reserved, so that the FPGA is initially empty, see also (3.14):

$$p_{i0k} = 0 \quad i \in \mathcal{S}, k \in U \tag{3.13}$$

Module reuse constraints A task can exploit module reuse only if in the time step preceding its arrival time an equivalent task occupies the same position:*

$$p_{ihk} + m_i \leq \sum_{j \in \mathcal{S}} a_{ij} p_{j(h-1)k} + 1 \quad i \in \mathcal{S}, h \in T \setminus \{0\}, k \in U \tag{3.14}$$

Definition of the total latency

$$t_e \geq S_i^{\text{off}} \quad i \in \mathcal{S} \tag{3.15}$$

3.3.2.4 Objective

$$\min t_e \tag{3.16}$$

3.3.3 Heuristic

3.3.3.1 Salomone

Salomone solves the scheduling problem trying to take advantages from the solution of the task allocation problem, that is, it assigns a location on the target device to each node of the input graph reconfiguring the functionality assigned to the same location. We will show the different phases involved using the TASK DEPENDENCE GRAPH shown in Figure 3.9 as an example.**

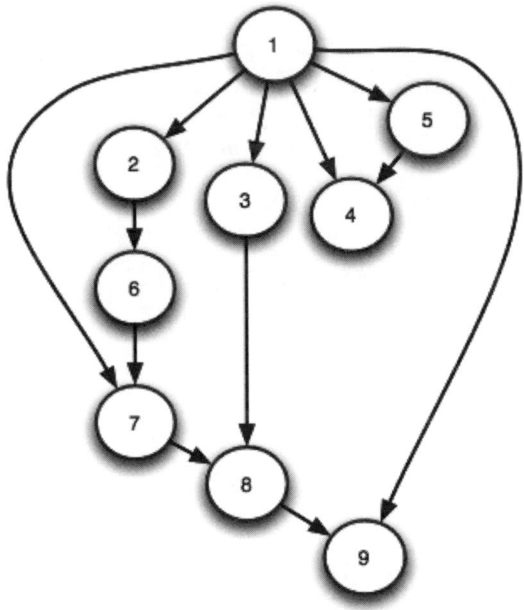

Figure 3.9: TDG from an ISCAS specification.

The basic idea is to try to identify a feasible location to each node, remembering that the switch between two nodes (two functionalities, two tasks) has to be implemented by introducing a time penalty due to the reconfiguration, which will provide useful information that, if used as input for the scheduler, will lead to the definition of a good scheduling solution.

The first phase aims at defining the conflict†† graph using as input the TDG.

**Which can be thought of as provided as input to the scheduling phase by the previous partitioning one.

††A conflict is modeled using an edge. It means that two nodes cannot be mapped on the same location since they can be executed in parallel.

This is done to identify the nodes able be executed in parallel. Starting from the same scheduling solution, more than one conflict graph can be obtained from the algorithm without increasing the computed solution. Experimental results proved that the best conflict graph is the one that has the smallest number of overlapping nodes, therefore this graph will be provided as output of this phase. At this point it is important to consider the following situation, which introduces an interesting distinction with respect to the classical problem, proposed in [41]. Consider two nodes S and S' having a data dependency $S \rightarrow S'$: it is desirable, since the reconfiguration time can introduce latency overhead, to have S' mapped on the chip as soon as S ends its computation. Therefore Salomone does not solve the allocation problem, using a coloring technique as done in the compiler design area [41, 40] on the TDG: rather, it considers a graph that is obtained merging the Conflict Graph and the Task Dependence Graph, the graph obtained from the task graph in Figure 3.9 is shown in Figure 3.10.

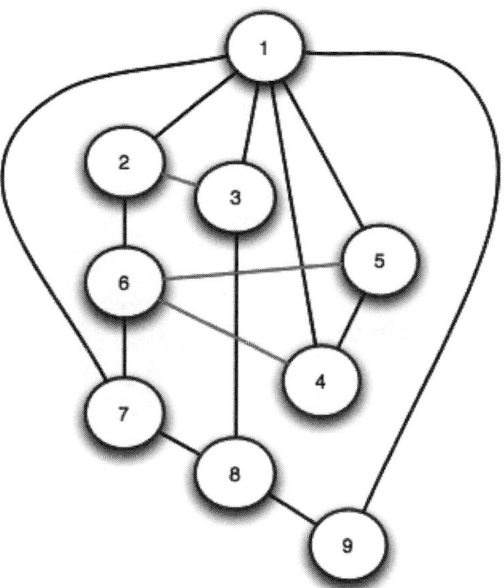

Figure 3.10: Task conflict graph.

Salomone then colors this new graph, partitioning it into several node sets, each of them corresponding to a specific color. This solves the allocation problem of each node onto the FPGA, partitioning the FPGA into a number of areas equal to the number of colors. This assignment will imply that each node belonging to a color set will be mapped onto the same FPGA area, partially reconfiguring the device, leading to define a SCONO (Same COlored NOdes) as an ordered set of nodes containing all the TDG nodes of the same color.

In order to color the graph Salomone uses `Adj`, the coloring algorithm proposed in [72, 66]. It has been designed to solve the coloring problem very rapidly while still retaining a good quality of the solution. The approach of the `Adj` algorithm is to scan all the nodes, coloring at each iteration not only the node v being considered, but also all its neighboring nodes $\texttt{Nb}(v) := \{w \in V | \exists (v, w) \in E\}$, unless they are already colored. If they are, it checks for color conflicts (i.e., two adjacent nodes having the same color). If during the scan of the neighbors of node v the algorithm finds that $w \in \texttt{Nb}(v)$ is colored and has a color conflict, it sets the color of node w to -1, and the conflict will be dealt with in the iteration considering node w. The `Adj` algorithm pseudo code is the following presented in Algorithms 7 and 10.

Algorithm 7 Adj

```
AllColors ← ∅;
for all (v ∈ V) do
  Colors[v] ← 0;
end for
for all (v ∈ V) do
  if Colors[v] = 0 then
    ColorThis(v);
    ColorNb(v);
  else
    if Colors[v] = −1 then
      ColorThis(v);
    else
      ColorNb(v)
    end if
  end if
end for
```

The choice of the actual color being assigned to a node is done via an heuristic based on the `Friend`$(c, Cols)$ function which, among the colors $Cols$, chooses the one that is most frequently adjacent to c in the graph at a particular time.

The `Friend` function checks how many edges exist with certain endpoint colors. Its complexity is therefore $O(|E|)$. The `CheckConflict` function, instead, has $O(|V|)$ worst-case complexity (for a dense graph). As for the `ColorThis` function, steps 1 and 2 require $O(|V|)$ while the else branch takes $O(|V|)$ plus the complexity of `Friend`. Hence we obtain $O(|V| + |E|)$. It is

Algorithm 8 Function ColorThis(v)

$AvailColors \leftarrow Colors \setminus \bigcup\{Colors[w] | w \in Nb(v)\}$;
if $AvailColors = \emptyset$ **then**
  $AllColors \leftarrow AllColors \cup \{|AllColors| + 1\}$;
  $Colors[i] \leftarrow |AllColors|$;
else
  $c \leftarrow \text{argmax}_c \, |\{w \in \text{Nb}(v) | Colors[w] = c\}|$;
  $Colors[v] \leftarrow \text{Friend}(c, AvailColors)$.
end if

then easy to see that `ColorNb` takes $O(|V|(|V| + |E|))$. From this it can be deduced that an upper bound for the worst complexity of the overall algorithm is $O(|V|^2(|V| + |E|))$. This bound is not tight at all, though, since it assumes that condition (a) in the main body is always true. Obviously this is not the case, since some coloring is performed by `ColorNb` executed for previous nodes. In the case of sparse graphs (and those arising from our problems always are), however, assuming that the cardinality of every neighborhood is bound by a constant, one gets $O(1)$ for `CheckConflict`, $O(|E|)$ for `ColorThis` and hence for `ColorNb`, so that the total complexity is $O(|E||V|)$.

Algorithm 9 Function ColorNb(v)

$c \leftarrow Friend(Colors[v], AllColors \setminus \{Colors[w] | w \in Nb(w)\})$;
for all $w \in \text{Nb}(v)$ **do**
  if $Colors[w] <> 0$ **then**
    ColorThis(w)
  else
    if CheckConflict(w) **then**
      $Colors[w] \leftarrow -1$
    end if
  end if
end for

Finally, what remains to be done is to schedule each SCONO according to the data dependency of the TDG, assigning different priorities to each node according to the critical path of the TDG. In order to do this Salomone computes the TDG critical path using a classical *Backflow Algorithm* [27], but it is possible to easily change this policy according to user needs. Having a local scheduler that can focus its attention on a small set of nodes improves execution time, thus decreasing the total time that the algorithm needs to terminate.

This phase produces a base line schedule used as input for an online sched-

Algorithm 10 Function CheckConflict(v)

$ForbiddenColors \leftarrow \bigcup\{Colors[w] | w \in \text{Nb}(v)\}$;
return $(ForbiddenColors \cap \{Colors[v]\} = \emptyset)$.

Algorithm 11 Function Friend(c, $Cols$)

if $Cols = \emptyset$ **then**
  $AllColors \leftarrow AllColors \cup \{|AllColors| + 1\}$;
  return $|AllColors|$;
else
  return $\text{argmax}_{f \in Cols} | \{w | w \in \text{Nb}(v),$
    $Colors[v] = c,\ Colors[w] = f\} |$.
end if

uler used to change the order in which the nodes are mapped on the FPGA according to the environment changes that can influence the performance of the system at runtime.

3.3.3.2 Napoleon

Salomone does not provide task relocation. Once a task has been assigned to a specific SCONO it cannot be relocated at runtime to a different position since the communication infrastructure is built on top of the SCONOs floorplanning assignment. In such a scenario the larger area constraint, where the area constraint for a SCONO is defined considering the most demanding task, that have to be implemented in that area, is assigned to each SCONO, with a considerable waste of resources. To avoid these problems, a new heuristic, named Napoleon, has been defined. This new solution tries also to maximize the task reuse without introducing reconfiguration overhead whenever possible.

Napoleon is a reconfiguration-aware scheduler for dynamically partially reconfigurable architectures. Its distinguishing features are the exploitation of *configuration prefetching*, *module reuse*, and *anti-fragmentation* techniques.

Algorithm 12 shows the pseudo-code of the proposed algorithm. First of all, Napoleon performs an infinite-resource scheduling in order to sort the task set $\mathcal{S}$ by increasing ALAP values. Then, it builds subset RN with all tasks having no predecessor. In the following, RN will be updated so as to include all tasks whose predecessors have all been already scheduled (*available tasks*).

As long as the dummy end task is unscheduled, the algorithm performs the following operations. First it scans the available tasks in increasing ALAP order to determine whether any of them can reuse the modules currently placed on the FPGA. Each time this occurs, task S is allocated to the compatible module p, which is the farthest from the center of the FPGA. This *farthest placement* criteria is an anti-fragmentation technique that aims at favoring future placements as it is usually easier to place large modules at the center of the FPGA [15]). The task is scheduled in the current time step t with module reuse and it is moved from the available nodes to the just scheduled ones (subset SN). When no further reuse is possible, Napoleon scans the available tasks in increasing ALAP order to determine whether any of them can be feasibly allocated to the FPGA in the current time step. The allocation is feasible when a sufficient number of adjacent columns are currently free or when they can be freed by removing a presently unused module. If this occurs, the position for task S is chosen once again by the *farthest placement* criteria. Unused modules can be present on the FPGA because Napoleon adopts *limited deconfiguration* as a second anti-fragmentation technique: all modules are left on the FPGA until other tasks require their space, in order to increase the probability of reuse. The task is scheduled in the current time step t without module reuse and it is moved from the available nodes to the just scheduled ones (subset SN). As soon as one task is scheduled, the scan is interrupted because of the following technological limitation: the scheduler can reconfigure a single task at a time. When all possible tasks have been scheduled, the set of available tasks RN is updated: Algorithm 13 does that by scanning the successors of the tasks in SN, which have just been scheduled, and determining the ones that must be added to RN. The schedule procedure is detailed in Algorithm 14. When module reuse occurs and the reconfiguration starting time is set to the current time step t. The execution starting time is tentatively set to t if reuse occurs, to $t + d_S$ if reuse does not occur and reconfiguration is required. Then, the procedure scans all predecessors of task S and possibly postpones the execution starting time to guarantee that all the predecessors of S have terminated. Therefore, there can be an interval between the end of the reconfiguration and the beginning of the execution of a task. This feature is denoted as *configuration prefetching*. Finally, the current time step is updated. To improve efficiency, instead of simply increasing it by one, Napoleon directly computes the next relevant time step by taking into account that the scheduler can reconfigure a single task at a time. For this reason, the next relevant time step follows the last time step in which the reconfiguration device is in use. Algorithm 12 does not report two tricks to increase efficiency. If in the current time step all configured modules are in use, the first **for** loop can be skipped, because reuse is presently not possible. If in the current time step there is not enough available area to place any task, the second **for** loop can be skipped, because no new placement is presently possible.

```
Algorithm 12 Algorithm Napoleon(𝒮,𝒫)
t ← 1
𝒮 ← computeALAPandSort(𝒮,𝒫)
RN ←findAvailableTasks(𝒮)
while S_e is unscheduled do
   SN ← ∅
   Reuse ← true
   for all S ∈ RN do
      p ← findFarthestCompatibleModule(S, t)
      if ∃p then
         schedule(S, t, p,Reuse)
         RN ← RN \ {S}
         SN ← SN ∪ {S}
      end if
   end for
   Reuse ← false
   for all S ∈ RN do
      p ← findFarthestAvailableSpace(S, t)
      if ∃p then
         schedule(S, t, p,Reuse)
         RN ← RN \ {S}
         SN ← SN ∪ {S}
         break
      end if
   end for
   RN ← RN∪ NewAvailableNodes(SN)
   t ← nextControlStep(t)
end while
return S_e.end
```

```
Algorithm 13 Function NewAvailableNodes(SN)
RN' ← ∅
for all S ∈ SN do
   for all S' ∈ successors(S) do
      if predecessors(S') are all scheduled then
         RN ← RN ∪ {S'}
      end if
   end for
end for
return RN'
```

Algorithm 14 Procedure Schedule(S, t, p,Reuse)

```
if Reuse = true then
  t_S ← t
else
  t̄_S ← t
  t_S ← t + d_S
end if
for all S' ∈ predecessors(S) do
  if t_S < t_S' + l_S' then
    t_S ← t_S' + l_S'
  end if
end for
```

Summary

This chapter proposes an insight into the advantages of partial dynamic reconfiguration. In order to formally describe the properties of the actual FPGAs, a model, general enough to be applicable to several different chips, but also not too abstract or simple, thus retaining the important features of the problem, is presented. The specification is given in terms of a DFG $\langle O, P\rangle$, where the node set O represents the *operations* and the arc set P represents the *precedence relations*: arc $o_i \rightarrow o_j$ indicates that operation $o_i \in O$ must terminate before operation $o_j \in O$ starts. Different partitioning approaches are described in Section 3.2, comparing them to identify the most promising one in achieving the best results using the partial reconfigurable capabilities. This chapter presents how to best exploit such capabilities, starting from the given specification, by defining specific solutions to the partitioning and the scheduling problem tailored for self-dynamically reconfigurable systems. The problem at hand will be described from the point of view of the actual physical architecture and then a novel model will be used to describe it in a precise mathematical way to allow its analysis. Finally, scheduling techniques are illustrated in Section 3.3.

Exercises and Problems

1. Differentiate between time and spatial partitioning.

2. Why can a regularity-driven approach provide benefits for the definition of a partial reconfigurable system?

3. Differentiate between the Napoleon and the Salomone approaches.
4. Describe how the communication between different cores can affect the identification of the reconfigurable cores.
5. Explain the reasons why the local isomorphism check has to be done.
6. Hierarchical template graph (HTG) and partial reconfiguration: list at least two benefits produced by the HTG during the runtime partial reconfiguration.
7. Try to extend the proposed ILP to support the 2D reconfiguration.

4

Operating System for Reconfigurable Systems

The added dimensionality of dynamic reconfiguration in a system has created the concept of a *hardware task*. Similar to software tasks, hardware tasks also need to be managed, scheduled, and placed in the reconfigurable logic. Hardware is no longer merely slave devices attached to the bus and controlled by software drivers. Hardware tasks are more active, just like software tasks. They may request synchronization primitives such as mutexes and semaphores. They may request bus and memory access. They may be created and killed dynamically. They may invoke each other, just like software function invocations. All these behaviors of hardware tasks must be managed by an *Operating System for Reconfigurable Systems* (OS4RS).

Though the concept of an OS4RS was proposed back in 2001 [205] by Wigley and Kearney, the actual implementation of a real OS4RS appeared much later, in 2004. Nevertheless, the state-of-the-art technology for OS4RS development is still in its infancy because there are still many hurdles to overcome before a full-fledged operating system can be developed that can meet all the desired requirements proposed initially in 2001. There are basically two types of OS4RS, namely target-specific and general-purpose. The target-specific OS4RS were built from scratch such as the MARC.1 OS4RS prototype [204, 206], the 3-tier ReConfigME OS4RS [208, 207], and the object-based management of NoC-connected hardware [83]. The general-purpose OS4RS were mainly extensions of the Linux OS such as the Egret/μCLinux OS4RS [209], IP-Core Manager (IPCM) [58], and BORPH [175, 176].

4.1 Motivation for OS4RS

Conventional operating systems (OS) manage software tasks, while hardware constitutes the lowest layer, mostly hidden by a hardware abstraction layer (HAL) along with software drivers. This segregation between hardware and software has benefited application development because it can follow a design flow independent of the underlying hardware architecture. However, in reconfigurable systems, where hardware is no longer slave devices, can actively request system resources and thus they will now contend with software tasks for the commonly required resources. This shows the need for the development of an operating system specifically for reconfigurable systems. In the following, we will further discuss in detail why an OS4RS is required.

First of all, a conventional OS manages only software tasks. There is no concept of a *hardware task* and task management requires kernel support, thus the management for hardware tasks is not possible in a conventional OS.

Second, in a reconfigurable system, hardware and software tasks need to communicate with each other at a peer level, which is unlike the conventional communication between the software running on a master CPU and the hardware of a slave device. Thus, a unified communication mechanism is required for hardware and software tasks.

Third, hardware tasks in a reconfigurable system need to be scheduled and placed; however, a conventional OS merely drives and controls the FPGA, with at most a software driver for the configuration controller, which is insufficient to manage hardware tasks.

Fourth, in a reconfigurable system, it is sometimes desired that a function can be implemented in both hardware and software such that either its hardware implementation or its software implementation is executed, depending on the system performance and application requirements. Extending this concept even further results in the so-called *task relocation*, where the execution of a task switches at run-time from its software implementation to its hardware implementation or vice versa. Managing task relocation requires the hardware and software tasks to have a uniform communication interface such that a task can continue communicating with other tasks even if it changes its implementation between software and hardware.

Finally, the configuration of a hardware design into the reconfigurable logic not only takes time, but also consumes power; thus, it becomes desirable that configurations are prefetched and reused. However, this requires an application level or module level management by the OS. Thus, an OS4RS is necessary.

4.2 Requirements for OS4RS

Corresponding to the various deficiencies in a conventional OS, an OS4RS would require covering them as described in the following, which includes introducing the concept of a hardware task, designing a hardware-software task communication mechanism, design of the scheduling and the placement algorithms for hardware tasks, hardware task reconfiguration and relocation, and a module manager that supports the reuse of hardware IPs.

First, the operating system needs to support the concept of hardware tasks, which could be very similar to software tasks, except their execution is performed in reconfigurable logic, instead of the microprocessor. However, several instances of hardware tasks could be executing in parallel, similar to software tasks running in parallel on multicore processors. In the BORPH OS4RS [175, 176], hardware tasks are encapsulated in `.bof` files that contain not only the hardware logic configuration data such as the Xilinx bitstream, but also task-related information.

Second, in an OS4RS it is desirable that hardware tasks communicate with software tasks seamlessly, that is, the communication mechanism does not depend on the implementation of the tasks. In other words, a task T_i communicating with another task T_j need not know if T_j is hardware or software. This can be achieved by adopting and extending the *interprocess communication* (IPC) mechanisms found traditionally in an OS for handling hardware and software task communication. A uniform task communication interface can be developed by a seamless integration of the hardware abstraction layer for reconfigurable logic with the hardware abstraction layer of a conventional OS.

Third, an OS4RS needs to manage not only software tasks, but also hardware tasks. Thus, a scheduler and a placer need to be developed for hardware tasks. The hardware task scheduler determines which ready tasks are to be selected for configuration and execution in the reconfigurable logic at different scheduling points. A scheduling point could be the time when a new task arrives or an existing task terminates execution. The hardware task placer manages the configurable resources in the reconfigurable logic. Since the configurable resources and the methods in which the configurable resources can be allocated for use depends heavily on the configuration architecture of the reconfigurable logic, the placer needs to take into consideration all the constraints of the logic device and its configuration capabilities.

Fourth, similar to the loading of software programs into random access memory (RAM), hardware designs also need to be *configured* into reconfigurable logic before they can be executed. Thus, an OS4RS needs to provide services through which hardware designs can be configured into reconfigurable logic. For example, a system call along low-level drivers for a configuration access port such as Xilinx ICAP (Internal Configuration Access Port) could be a possible solution. A more advanced function of an OS4RS could also be *task relocation*, which allows an already placed and configured hardware task to be relocated into another area of the reconfigurable logic space, that is, using another set of resources to continue performing the same function from where it was preempted.

Finally, similar to a conventional OS, the overheads of an OS4RS should be minimized such that applications are efficiently executed. However, the management of tasks inevitably consumes time and power because software code of the OS4RS needs to be executed. To amortize this overhead, an efficient reuse of hardware IP designs across different applications can be supported by an OS4RS.

In summary, an OS4RS should at least provide the following services: scheduling, placing, and loading a reconfigurable hardware task and a uniform interface for communication between hardware and software tasks. Advanced services of an OS4RS could include dynamically partitioning a task as hardware or software, relocating a hardware task as a software one and vice versa [134], de-fragmentation of the reconfigurable logic resource space, module management such that hardware and software implementations of a task are reused across different applications, support for *Network-on-Chips* (NoC) [131, 140].

Some optional but desirable features of an OS4RS could include portability across different architectures of reconfigurable logic and the provisions for a threaded model of hardware tasks such that a *hardware thread* can use the same communication interfaces as the software tasks. A typical example is the *Hybridthread* (Hthread) model for hardware tasks as proposed by Anderson et al. [7]. Hthreads was proposed to allow a programmer to write a multi-threaded pthreads program, compile it, and have it run efficiently on a hybrid CPU/FPGA chip. With Hthreads, a thread could run as either a software thread on the CPU or as a hardware thread in the FPGA. The programmer need only be concerned with program correctness, thus Hthreads help free him or her from the technical details of hardware design.

4.3 Layered Architecture for OS4RS

From the previous discussions, we can clearly see that an OS4RS needs to provide services at different levels of the system hierarchy. Thus, a layered architecture is generally proposed for an OS4RS. Another reason is that the metamorphosis of tasks between its hardware and software implementations requires a layered approach such that application level, task level, and function level attributes and characteristics are clearly distinguished through the different layers.

As illustrated in Figure 4.1, the layered architecture of an OS4RS consists of six layers, namely hardware, configuration, placement, scheduling, module, and application layers. We will now describe each of these layers in more detail. Note that some OS4RS architectures try to integrate the layers into a single *Hardware Abstraction Layer* (HAL) [134] or into three tiers including the user tier, OS tier, and platform tier [207, 208]. A larger number of layers allows a modular and hierarchical development of the operating system; however, it might affect the efficiency by introducing overheads. We need to ensure that such overheads are minimized in the layered architecture.

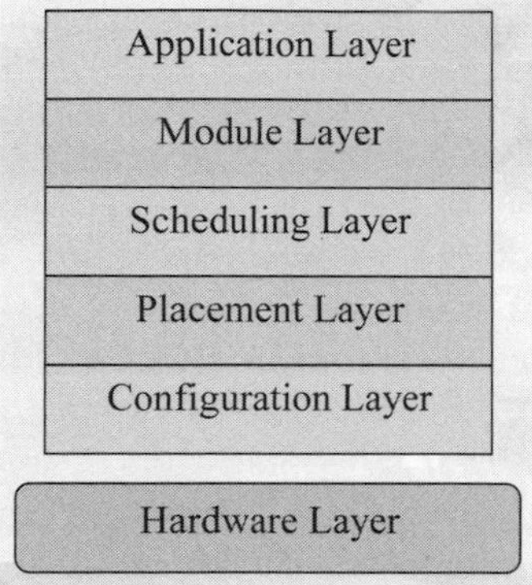

Figure 4.1: Layered architecture of OS4RS.

4.3.1 Hardware Layer

The lower most layer is the hardware layer, which consists of at least a fine-grained reconfigurable logic such as FPGA and a general-purpose microprocessor such as PowerPC or ARM processors, which are responsible for the running the hardware and the software tasks, respectively. However, for a feasible reconfigurable system design, an efficient system architecture that can support an OS4RS is desired. Currently, there is no *standard* architecture for dynamically reconfigurable systems. A basic reference architecture is shown in Figure 4.2, which consists of a processor for executing software tasks, a memory for data storage, a hardware accelerator that represents one or more fixed hardware designs, a reconfigurable logic that can be configured at run-time and partially for executing some hardware functions, a function ROM for storing the configuration data such as Xilinx bitstreams, and a shared bus that connects all the system components to allow memory access and data communication.

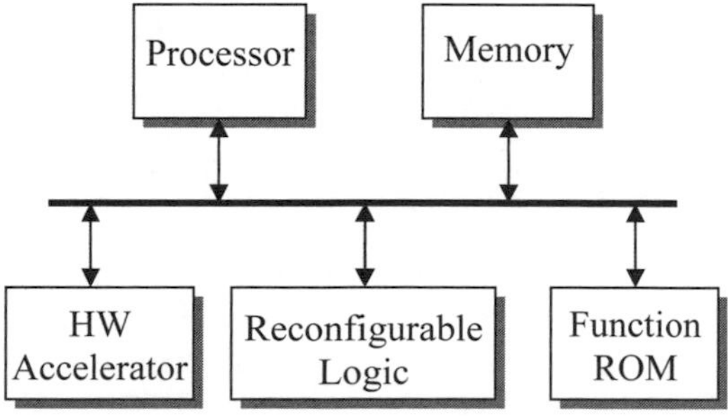

Figure 4.2: Dynamically reconfigurable system architecture.

The reference architecture can be easily extended with more advanced features as described in the following.

- Network-on-Chip (NoC): NoC is a more efficient communication infrastructure that can allow multiple concurrent data transfers using different switching and routing schemes. Reconfigurable NoC is also currently a hot topic of research because it not only makes the NoC design more adaptable to dynamic application changes, but also brings unforeseen advantages to the traditional NoC [168] such as run-time migration of processing elements (PE), de-allocation of unused PE, dynamic mapping of PE to the available NoC routers, replacing a simple (complex) router with another complex (simple) router based on the needs of the applications, shutting down part of the NoC and restarting it when and where required, reconfiguration-based congestion avoidance strategies, predictable data transfers, and reduction in power consumption, to name a few. The OS4RS will play an important role in such a reconfigurable NoC due to the added features.

- Reconfigurable Module Sequencer (RMS): RMS [169] is more like a hardware task manager, but its main job is to alleviate the burden of the microprocessor and the OS4RS in handling the data communications among different hardware tasks. System calls are implemented in the OS4RS to drive the RMS. A conventional program must be first transformed into a chained program, which has a set of chains that can be executed by the RMS. A chain is a sequence of reconfigurable hardware functions. The data transfers between consecutive hardware functions in a chain are handled by the RMS. Thus, the communication performance of a reconfigurable system with RMS is higher than that without RMS.

- Task Configuration Controller (TCC): The basic function of a task configuration controller is to read configuration bitstreams and send them to a configuration port such as ICAP. It could be implemented in hardware or software depending on the availability of hardware computing resources and real-time constraints. More advanced functionalities of a TCC could include task context saving and restoring, task switching, and task relocation. Details of these advanced functions can be found in Section 4.3.5. To support TCC, an OS4RS must include related device drivers such as ICAP driver and control programs such as the *mediator* program [180].

- Data Sequencer: A major performance bottleneck for reconfigurable hardware tasks is data streaming. The hardware tasks are usually slaves and thus data need to be streamed into and out of the tasks. To solve this problem, either the hardware tasks should be implemented as master devices or the system architecture could include a data sequencer. Master hardware tasks are similar to the Hthreads concept

[7], where a hardware thread can access shared IPC structures, synchronization primitives, and perform I/O just like software threads. A data sequencer can be used to accelerate data I/O to and from the hardware tasks through a devoted communication infrastructure such as a bus or NoC.

4.3.2 Configuration Layer

The configuration layer is responsible for all functions related to the (re)configuration of hardware tasks into the reconfigurable logic. The main components in this layer include a configuration access port and a configuration controller. The configuration access port is used by a configuration controller to reconfigure parts of a chip logic. A typical example is the Xilinx *Internal Access Control Port* (ICAP) and its software driver, which is the configuration controller. The ICAP driver must be part of the OS4RS kernel.

Optional components in this layer could include a task context switch controller and a task relocation filter. A task context switch controller is responsible for saving and restoring the data context and state information of a reconfigurable hardware task [180]. This controller could be implemented in software and embedded within the OS4RS kernel. A task relocation filter is responsible for relocating a task from some source position to a destination position in a reconfigurable logic. In a 1-dimensional reconfiguration architecture such as that found in Xilinx Virtex II Pro, the task position is specified by the column number, whereas in a 2-dimensional one such as Xilinx Virtex 4 and 5, it is specified by the tile number and the column number. A tile is a set of CLB rows (16 rows in Virtex 4 and 20 rows in Virtex 5). The number of columns in a tile depends on the FPGA chip size. Similar to the switch controller, a relocation filter is also embedded within the OS4RS kernel.

4.3.3 Placement Layer

The placement layer consists of a placer that manages the reconfigurable resources by allocating or de-allocating them to hardware tasks. The basic unit of allocation could be a column of *Configurable Logic Blocks* (CLBs) that spans the full height of a chip as in Xilinx Virtex II Pro or it could be a column of CLBs that spans the height of a tile, where a tile consists of several columns of fixed number of CLBs such as 16 CLB columns in Xilinx Virtex 4 and 20 CLB columns in Xilinx Virtex 5. We will henceforth call this the *Basic Unit of Configuration* (BUC).

Due to the limitation of currently available reconfigurable devices, hardware configurations must be constrained in rectangular areas. Based on this constraint and the basic unit of configuration, three *reconfigurable area models* are generally used, as illustrated in Figure 4.3 and described in the following.

- *Slotted Area Model*: In the slotted area model, as shown in Figure 4.3(a),

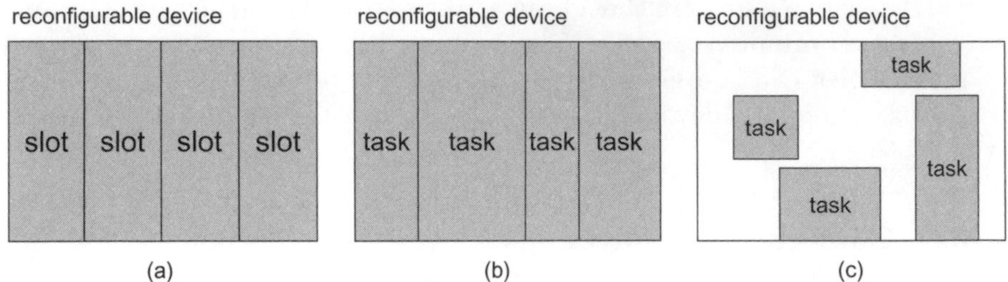

Figure 4.3: Reconfigurable area models.

the reconfigurable device is divided into a number of slots, each of which is of the same size. Hardware tasks are synthesized by design tools to occupy one or more slots. These tasks can be relocated to any contiguous set of slots in the reconfigurable area of a device. The slotted area model has the problem of *internal fragmentation*, that is, if the area occupied by a task is smaller than the allocated slot area, the remaining free area is wasted. The slotted area model was used in [133, 139] for hardware task placement.

- *1D Area Model*: In the 1D area model, as shown in Figure 4.3(b), the vertical dimension is fixed and spans the height of the device. Tasks can be allocated anywhere along the horizontal dimension with no preallocated margins as in the slotted area model. The 1D area model suffers from both internal and *external fragmentation*, where external fragmentation means the overall free area may be bigger than the area required by a task; however, the free area is not contiguous such that it becomes impossible to place the task. The 1D area model was used in [29] for hardware task placement.

- *2D Area Model*: The 2D area model, as shown in Figure 4.3(c), is the most flexible model that allows us to allocate tasks anywhere on the device. The advantage of this model is high device utilization. However, the high flexibility of this model makes scheduling and placement more difficult. The 2D area model has the problem of *external fragmentation*, that is, if tasks are placed at arbitrary positions the remaining free area is fragmented. The 2D area model was used in [20, 28, 55].

A reconfigurable hardware task is allocated resources in multiples of the CLB columns, which might result in both *internal* and *external* fragmentations of the available configurable resources. This is very similar to the internal and external fragmentations that occur in paged memory allocation schemes. Memory is allocated in fixed units called pages, where the unused memory within a page results in internal fragmentation and the scattering of free pages results in external memory.

Internal fragmentation results from a hardware design not completely using the CLBs in the allocated columns. This situation is very evident in Xilinx Virtex II Pro FPGA devices because the columns span the full chip height, though a simple hardware design such as a half-adder might use only a few CLBs in the column. The rest of the CLBs are unused and wasted until the hardware task completes and resources are released. The issue of internal fragmentation has been alleviated in Virtex 4 and 5 because of two reasons, including: (1) each column is now much shorter and consists of only 16 or 20 CLBs, which results in lesser wastage, and (2) under some restrictions the CLBs unused by a reconfigurable hardware design can still be used by the static system hardware, thus they are not completely wasted. The restrictions are device dependent and at the time of writing this book, in Virtex 4 and 5 FPGAs, the static hardware must not configure the unused CLBs as LUTRAM (memory) or SRL16 (shift logic registers).

External fragmentation occurs when the free columns are scattered throughout the chip area such that there are very few contiguous columns that are big enough to accommodate a reconfigurable hardware design. To solve the external fragmentation problem, different approaches have been proposed in the literature, including the adjacency-based heuristic and the fragmentation-based heuristic [181].

Besides minimizing fragmentation of available resources, other goals of placement algorithms include the minimization of routing resources and the minimization of the time required for computing a placement location. We will now describe some representative methods for each of these goals and also present a multi-objective hardware placement strategy in the rest of this section.

4.3.3.1 Minimizing Fragmentation

The placement methods that focus on reducing fragmentation are described briefly in the following.

- *Adjacency Heuristic*: Among all possible locations that can accommodate a newly arrived task, the adjacency heuristic chooses a location that has the highest contact level between the new task and the adjacent running task edges or the device boundaries. The contact level is counted in terms of the basic unit of configuration such as Xilinx FPGA column. As an example, consider the task placement situation depicted in Figure 4.4, where tasks A, B, and C are already placed and we need to place task D in the available area. The newly arrived task D can be accommodated in each of the 6 dotted rectangular areas. However, the contact level is maximum for the top left corner area, thus task D will be placed there by the adjacency heuristic.

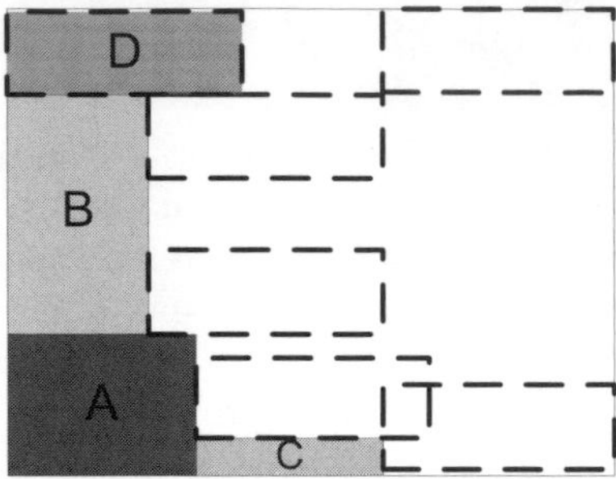

Figure 4.4: Adjacency heuristic.

- *Fragmentation Heuristic*: Given a reconfigurable logic of area A_F and a set of free spaces with area A_i that can be represented by V_i vertices, the fragmentation heuristic defines a fragmentation metric as follows:

$$F = 1 - \Pi_i[(4/V_i) \times (A_i/A_F)] \tag{4.1}$$

Note that $V_i \geq 4$ because the rectangular area represented by 4 vertices is the most regular one and hence causes the least fragmentation. Also note that a smaller value is desirable for F. An example is illustrated in Figure 4.5, where tasks A, B, and C are already placed and we need to place task D in the available area. Before placing D, there is only one free space represented by 10 vertices. Each of the 6 dotted rectangles is a candidate location. However, note that only the locations at the lower right corner and the upper left corner result in a free space that is still representable by 10 vertices. All other candidates result in more than 10 vertices, that is, $V_i > 10$ and thus a higher fragmentation F. The fragmentation heuristic places task D at either the lower right corner or the upper left corner.

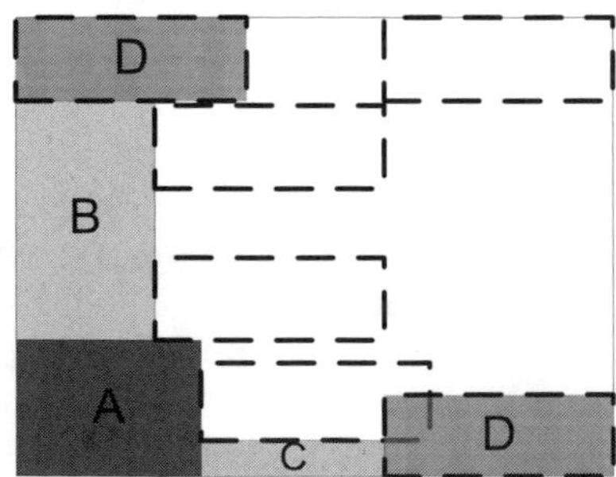

Figure 4.5: Fragmentation heuristic.

- *Best/Worst Fit with Exact Fit Heuristic*: Among all free spaces that can accommodate a new task, the best (worst) fit with exact fit heuristic [201] chooses the smallest (biggest) rectangle that has exactly the same width or exactly the same height as the task. If no such rectangle is found, the task will be placed according to the conventional best (worst) fit strategy.

From the above fragmentation minimization heuristics, one can see that the best/worst fit with exact fit heuristic is the simplest but might not reduce fragmentation as much as the other heuristics. Further, the adjacency heuristic is simpler than the fragmentation heuristic; however, it might not provide as many feasible solutions as the fragmentation heuristic. Nevertheless, the fragmentation heuristic requires more computations. Irrespective of the heuristic, the placement algorithm complexity is $O(n^2)$, where n is the number of running tasks.

4.3.3.2 Minimizing Routing Resources

Besides reducing fragmentation, placement methods for reconfigurable hardware tasks also have other considerations such as reducing the routing resources for communicating tasks. For independent tasks that do not communicate with one another, there is no need to consider routing resources. However, for communicating tasks, routing resources are required to link them together, but the amount and the density of routing resources available in a reconfigurable logic architecture are statically fixed at design time. It is observed that the longer the distance between the two communicating tasks, the more routing resources are required to link them together. Not only is the routing cost increased due to inter-task distance, but the data communication latency between the tasks is also increased, thus affecting the system performance. Hence, it is preferable to minimize the routing distance between communicating tasks.

Suppose that tasks T_i and T_j communicate with each other. Note that either of the tasks could represent output pins on the device boundaries, that is, a task could need to connect to the device boundaries for I/O device access. Let (w_i, h_i) and (w_j, h_j) be the width and height of the two tasks T_i and T_j, respectively. Further, let the tasks T_i and T_j be placed at locations with bottom left positions (x_i, y_i) and (x_j, y_j), respectively. Let the required communication bandwidth between T_i and T_j be w_{ij} in bits. If there is no communication between T_i and T_j, w_{ij} is zero. Ahmadinia et al. [3, 4] defined the routing cost between two tasks as the square of the distance between the centers of the the two rectangles representing the two tasks, multiplied by the bandwidth. Notationally, the routing cost between two tasks T_i and T_j is formally defined as follows:

$$RCost(i,j) = ((\Delta_x(i,j))^2 + (\Delta_y(i,j))^2) \times w_{ij} \tag{4.2}$$

where $\Delta_x(i,j) = (x_j + w_j/2 - x_i - w_i/2)$ and $\Delta_y(i,j) = (y_j + h_j/2 - y_i - h_i/2)$ are the differences in the x-coordinates and the y-coordinates, respectively, of the centers of the two rectangles representing the two tasks.

Suppose $n-1$ tasks are already placed in a reconfigurable logic area. We need to find a suitable location (x_n, y_n) for an incoming task T_n such that the total amount of routing resources is minimized. This objective of the placement strategy can be formalized as Equation 4.3. To achieve this objective, the optimal location (x_n, y_n) can be calculated as derived in Equations 4.4, 4.5, and 4.6. The placement algorithm computes in $O(n)$ time, where n is the number of running tasks.

$$TotalRCost(i,n) = \min_{x_n, y_n} \{\sum_{i=1}^{n-1} [((\Delta_x(i,n))^2 + (\Delta_y(i,n))^2) \times w_{in}]\} \tag{4.3}$$

$$\frac{\partial\{\sum_{i=1}^{n-1}[((\Delta_x(i,n))^2 + (\Delta_y(i,n))^2) \times w_{in}]\}}{\partial x_n} = 0 \tag{4.4}$$

$$\frac{\partial\{\sum_{i=1}^{n-1}[((\Delta_x(i,n))^2 + (\Delta_y(i,n))^2) \times w_{in}]\}}{\partial y_n} = 0 \tag{4.5}$$

$$x_n = \frac{\sum_{i=1}^{n-1} w_{in} \times (x_i + w_i/2 - w_n/2)}{\sum_{i=1}^{n-1} w_{in}} \tag{4.6}$$

$$y_n = \frac{\sum_{i=1}^{n-1} w_{in} \times (y_i + h_i/2 - h_n/2)}{\sum_{i=1}^{n-1} w_{in}} \tag{4.7}$$

Take Figure 4.6 as an example to illustrate the minimization of routing cost placement strategy. Similar to the previous examples, tasks A, B, and C are already placed and we would like to place task D now. Depending on the communication bandwidths w_{AD}, w_{BD}, and w_{CD}, any of the two closely placed blocks could be candidate locations for placement.

However, the computed location (x_n, y_n) could be infeasible for placement if the selected area overlaps with one or more tasks that are already placed. In this case, another heuristic method called *Nearest Possible Positions Compute* (NPP-Compute) [3, 4] can be used to find the feasible nearest possible positions outside the *Impossible Placement Regions* (IPR) of all tasks. Corresponding to an incoming task D, we can compute IPR(X) by considering all the positions surrounding task X in which task D cannot be placed. An example on how the NPP-Compute heuristic works is illustrated in Figure 4.7, where the optimal position is infeasible because it overlaps with task E, which is already placed. Starting from the optimal position and avoiding all the impossible placement regions IPR(A), IPR(B), IPR(C), and IPR(E), NPP-Compute finds the nearest possible position where task D can be placed as shown in Figure 4.7.

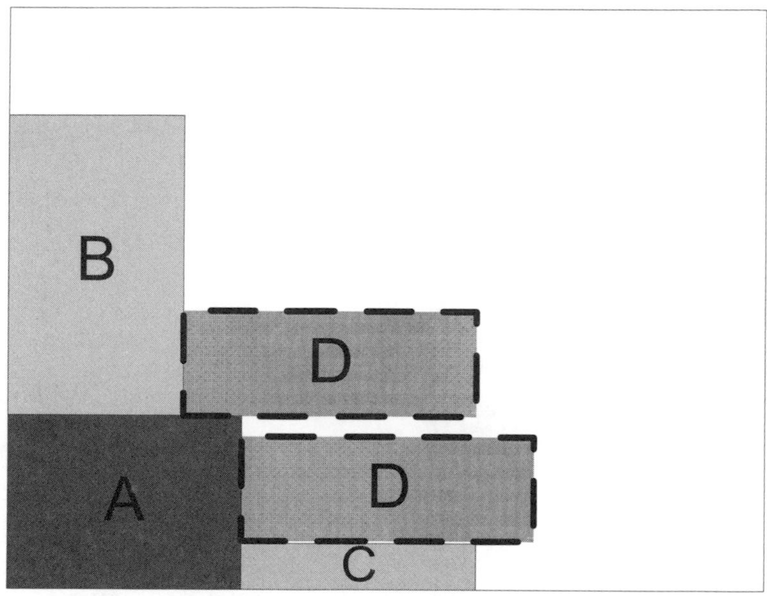

Figure 4.6: Minimization of routing resources.

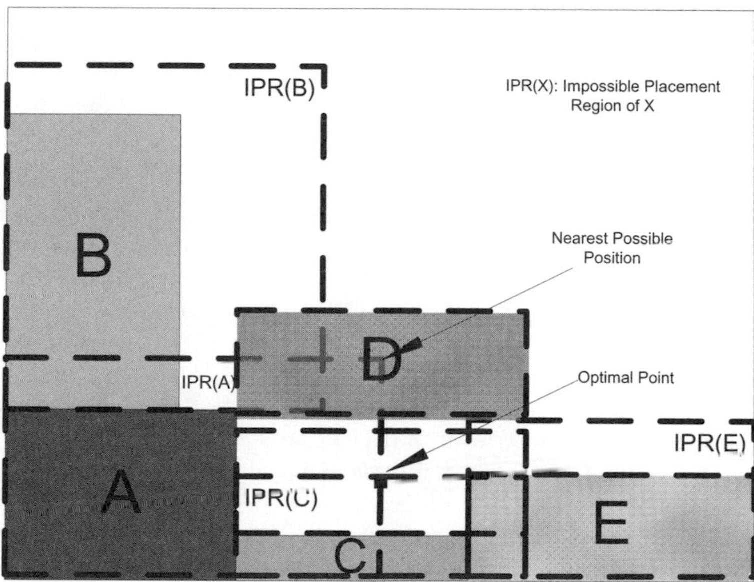

Figure 4.7: Nearest possible position selection in minimization of routing strategy.

4.3.3.3 Fast Placements

A major bottleneck in hardware task placement methods is the extensive computations required in evaluating each and every possible location to place a task. With rapidly increasing sizes of reconfigurable logic such as the Xilinx Virtex 5 and Altera FPGA devices, the number of possible locations becomes even more unmanageable. Additionally, a newly arrived task might have very tight deadlines, which make such time-consuming computations infeasible. Fast placement strategies try to find a possible location within a restricted amount of time. Well-known strategies include the *first fit* (FF), the *best fit* (BF), and the *bottom left* (BL) strategies [20], which are similar to the memory allocation strategies. A first fit strategy chooses the first free space that can accommodate the hardware size of the new task. A best fit strategy chooses the smallest free space that can accommodate the hardware size of the new task. The bottom left strategy tries to place a new task in a sufficiently big free space whose lower left corner is located closest to the lower left corner of the reconfigurable device. All of these algorithms are linear in time complexity in terms of the number of free spaces.

Configuration overhead is a major limitation in most fine-grained reconfigurable logic devices. Besides using scheduling-based methods such as configuration prefetch and configuration reuse, placement-based methods also exist to reduce configuration overhead. The *Least Interference Fit* (LIF) [5] chooses the free space that has the least number of already placed tasks under the free space. The intuition is that for column-based reconfigurations in some older Xilinx FPGAs such as Virtex II Pro, whenever we are configuring some part of a column all other placed tasks occupying the same column must be suspended.

4.3.3.4 Multi-Objective Hardware Placement

Most placement methods have a single goal such as one of those described above, namely minimization of fragmentation, minimization of routing resources, and minimization of time for location selection. However, in a real system, it is often required to satisfy multiple goals at the same time. *Multi-Objective Hardware Placement* (MOHP) [124] is one of the methods to achieve this.

MOHP is simply an integration of three single-objective placement methods with the three goals as described above. Tasks are classified into three types, including (a) communicating, (b) independent and non-real-time, and (c) independent and real-time. The different types of tasks have different goals. Communicating tasks are placed such that the average routing resource usage is minimized. Independent non-real-time tasks are placed such that fragmentation is minimized. Independent real-time tasks are placed such that location selection is fast, namely the first-fit strategy.

MOHP employs different starting locations for the different types of tasks. The minimization of routing resources (MR) algorithm starts placing tasks from the upper right corner of a reconfigurable logic area. The minimization of fragmentation (MF) algorithm starts from the lower right corner and the first-fit (FF) algorithm starts from the lower left corner. If a fourth placement algorithm is to be incorporated into MOHP, the upper left corner can then be used. Given a large enough reconfigurable logic device such as Virtex 5 or Altera Stratix III, effective separation of the different types of tasks allows the simultaneous satisfaction of the different goals.

The flow chart for the MOHP algorithm is illustrated in Figure 4.8. MOHP employs three queues to schedule the different types of tasks that are ready to be placed and executed. The routing queue consists of the communicating tasks, the fragment queue consists of the independent non-real-time tasks, and the urgent queue consists of the independent real-time tasks. The MR, MF, and FF algorithms are used for placing the tasks from the routing queue, the fragment queue, and the urgent queue, respectively. If a feasible location is found by the algorithms, then the hardware task is placed. Otherwise, the task waits for the next chance, which could be the next scheduling point such as when a hardware task terminates.

4.3.3.5 Resource Management Strategies

Along with several resource allocations to tasks and de-allocations after tasks finish, the available free space is fragmented into several areas and must be recorded and updated continuously. Furthermore, with the increase in the sizes of reconfigurable logic, this maintenance of data structures becomes a time-consuming computation in placement methods because with each placement, all the different data structures such as the task list, the free space list, and the reserved space list must be updated. There are several different ways to maintain the records and to accelerate the updating of the records [3, 4, 20, 79, 80, 179, 181, 182, 201].

Bazargan et al. [20] proposed *Maximal Empty Rectangles* (MER) to record the free spaces, where an MER is a rectangular area that is not contained within any other empty rectangle. Using MER, it can be guaranteed that if there is an adequate space for an incoming task, it can be found, but at the cost of a high complexity. Another method cuts the empty region into several *Non-Overlapping Empty Rectangles* (NER), which are not necessarily maximal in area. These free rectangles are candidates for accommodating an incoming task. The favored location for an incoming task is based on the bin-packing rule such as FF or BF. Finding maximal rectangles takes quadratic space in terms of the number of running tasks, while finding non-overlapping rectangles needs linear space. Some placement quality is lost when using NER, but the placement algorithm is faster and easier to implement.

Steiger et al. [179] [201] proposed an efficient partitioner, which enhanced Bazargan's partitioner [20] by delaying the space partition and using a 2D hash structure to maintain the empty spaces. The 2D hash structure allowed a placer to find free spaces more quickly. Thus, the placer could locate real-

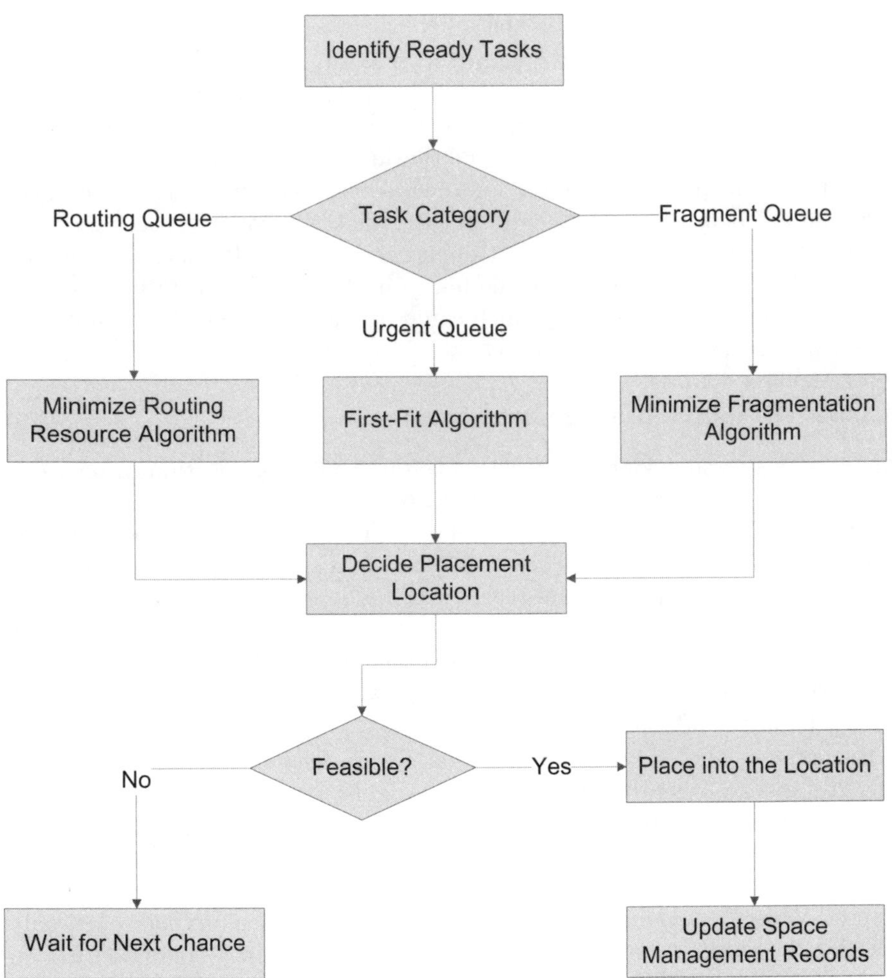

Figure 4.8: Multi-objective hardware placement.

time tasks quickly.

Handa and Vemuri [79] presented an algorithm that aims at finding empty space fast and efficiently. They made use of two data structures, namely an *area matrix* to model the FPGA resources and a *staircase data structure* to model an empty space in the form of a list of MERs. In an area matrix representation, each cell of the matrix was assigned a negative value for occupied space and a positive integer number for empty space. The search for a staircase data structure has a complexity of $O(m \times n)$, where m and n are the number of columns and rows, respectively. Handa and Vemuri [80] also developed three kinds of space managers with hardware implementations, including three architectures, namely serial, parallel, and serial-parallel architectures.

Tabero et al. [181] managed the free spaces in a reconfigurable device using a *Vertex List Set* (VLS) such that a free space fragment is represented by a list of vertices at the boundary of the fragment. They achieved the goal of minimizing fragmentation using two heuristics, namely the adjacency-based heuristic and the fragmentation-based heuristic, as described earlier. Later, Tabero et al. extended the 2-dimensional model in the 3-dimensional space management model with an extra time axis [182].

All the above described placers search for a suitable free space for placement by maintaining the empty spaces; however, Ahmadinia proposed the management of occupied spaces instead [3, 4]. Ahmadinia et al. found that the records of the occupied spaces grow more slowly than those for free spaces, which means that the data preserved for occupied spaces are smaller in size than that for empty spaces. The authors utilized this characteristic of space managers, and they aimed at optimal inter-task communications, which consisted of not only connections between tasks and boundary of devices, but also connections between placed tasks.

4.3.4 Scheduling Layer

Going one layer above the placement layer is the scheduling layer, which is responsible for deciding the start time of reconfigurable hardware task executions. However, the decision of when to start a task execution depends on whether the task can be placed and configured into the reconfigurable logic. Thus, scheduling and placement are intricately related in reconfigurable systems, which is different from the task scheduling and memory allocation that can be separately performed in conventional systems. For example, in a conventional system, a task is first loaded into memory and then it must wait in the ready queue for its turn to use the CPU. In reconfigurable systems, as soon as a task is configured into the reconfigurable logic, it can begin execution because configuration not only means task loading but also the allocation of processing resources. Nevertheless, we will first describe the application of conventional scheduling methods to hardware task scheduling and then more specific methods targeted at hardware task scheduling. Finally, we will describe methods that have additional goals such as the reduction of fragmentation, the reduction of energy consumption, and the increasing of hardware utilization.

In general, scheduling methods can be classified into static scheduling and dynamic scheduling. A static scheduler is employed at design time, while a dynamic scheduler is used at runtime to schedule the dynamically arrived tasks. The static scheduler uses statically known information, such as the task arrival times and the task execution times, to decide the task execution sequence. It is suitable only for systems whose behavior are known in advance such as embedded systems that have fixed functionalities.

Most static schedulers for reconfigurable systems are based on the list scheduling method and use a queue to keep tasks that are ready to run. Given a list of ready tasks sorted by priorities, the scheduler selects the task with the highest priority to execute. The difference between these schedulers is the method for calculating the task priority. Some typical ways to assign task priority include random assignment, *As Soon As Possible* (ASAP), and *As Late As Possible* (ALAP). ASAP schedules a task to be executed as soon as possible, while ALAP schedules a task to be executed as late as possible.

Loo and Wells [127] introduced a high-level model called *Reconfigurable System Design Model* (RSDM) as illustrated in Figure 4.9. The scheduler in the RSDM model takes resource constraints, a hardware library consisting of processor cores (PCs), task specific cores (TSCs), and communication core elements (CCEs), a set of system tasks, and design constraints as input. The outputs of the scheduler include a feasible task schedule and a high-level hardware system description. RSDM used *random priority list scheduling* along with *simulated annealing* (SA) and *genetic algorithm* (GA) to schedule reconfigurable hardware tasks. As we know, SA and GA are well-known optimization algorithms that produce near-optimal results, but they are quite time-consuming and hence not suitable for application at runtime.

However, the behavior of reconfigurable systems changes dynamically and the static scheduler is not suitable for dealing with dynamic system behavior. Thus, a dynamic scheduler is required to schedule the dynamically arrived tasks. Due to the strict timing requirements at runtime, dynamic scheduling must be efficient such that a good trade-off is achieved between the quality of scheduling results and the impact of scheduling overhead on the system performance.

4.3.4.1 Adopting Conventional Scheduling Methods

Most dynamic scheduling algorithms used in a conventional operating system can be adopted for hardware task scheduling, as long as we do separate the placement considerations. However, as mentioned earlier in this section, scheduling and placement go hand in hand for reconfigurable systems. A situation might occur such that according to some conventional scheduling method task t_1 must be scheduled and executed before task t_2; however, the size of t_1 cannot be currently accommodated by the available reconfigurable

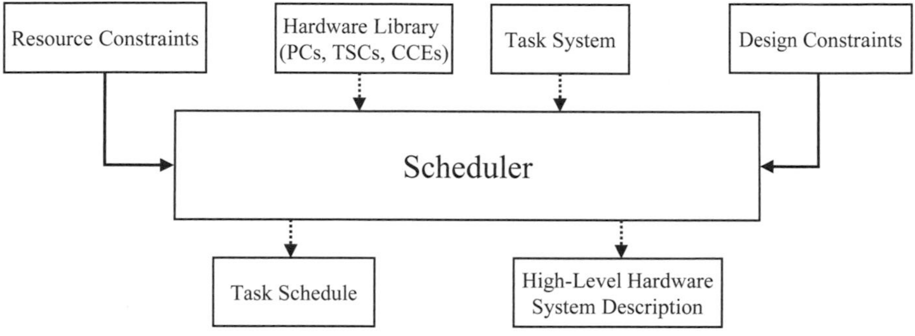

Figure 4.9: Reconfigurable system design model.

logic resources, while that of t_2 can be accommodated. A question arises as to whether we should schedule task t_2 first such that the hardware resource utilization is increased, which could be one of our scheduling goals. However, this disrupts the scheduling sequence decided by the selected conventional scheduling method.

Some conventional scheduling methods that have been used by Walder and Platzner [199] in an online scheduler for block-partitioned reconfigurable devices include *First Come First Serve* (FCFS), *Shortest Job First* (SJF), *Shortest Remaining Processing Time* (SRPT), and *Earliest Deadline First* (EDF). FCFS and SJF are non-preemptive methods, while SRPT and EDF are preemptive. In FCFS, tasks are sorted by their arrival times, and the scheduler will select the earliest arriving task to run. The SJF scheduler will sort the queue according to the execution time of the tasks, and the task with the shortest execution time will be selected first. The SRPT scheduler sorts the queue according to the remaining execution times of tasks, and picks the task with the shortest remaining execution time to execute first. EDF scheduler sorts tasks by their deadlines, and the task with the earliest deadline will be selected first.

A dynamic version of the ASAP/ALAP scheduling called *dynamic priority list scheduling* was also proposed or used for reconfigurable systems [132], which combines ALAP and dynamic ASAP to calculate task priority. Dynamic ASAP computes task priority at run time, and all dynamic ASAP values are recalculated once a task is scheduled. Thus task priorities are changed dynamically.

4.3.4.2 Reconfigurable Hardware Task Scheduling

Besides adopting conventional scheduling strategies, there are also several scheduling algorithms proposed specifically for reconfigurable hardware tasks. Similar to placement methods, the scheduling methods also have different goals, including the improvement of hardware resource utilization [49], the reduction of reconfiguration overhead [156, 155, 154], the reduction of scheduling overhead [177, 178], the reduction of fragmentation [42], and the reduction of power consumption [126, 137]. There is also a class of co-scheduling algorithms for hardware software tasks [43, 149] that can be relocated, where relocation means a hardware task can be preempted and restarted as a software task, while a software task can be preempted and restarted as a hardware task.

4.3.4.2.1 To improve hardware resource utilization Danne and Platzner proposed two preemptive scheduling algorithms [49] for reconfigurable hardware tasks such that the hardware resource utilization can be improved. One of the methods is an adaptation of the EDF method to the FPGA execution model, called *EDF-Next Fit* (EDF-NF). If the task with the earliest deadline can be configured then EDF-NF reduces to EDF. If the task with the earliest deadline cannot be placed due to lack of reconfigurable hardware resources, then EDF-NF continues to schedule tasks that can be placed into the available resources. Thus, EDF-NF can improve hardware resource utilization. The EDF-NF algorithm has good scheduling performance, but this approach is practical only for a small number of tasks. For larger task sets, another method called the *Merge-Server Distribute Load* (MSDL) algorithm was proposed. The MSDL algorithm uses the concept of servers that reserve area and execution time for tasks, thus it is suitable for large sets of tasks. However, MSDL is not as efficient in improving hardware resource utilization as EDF-NF.

4.3.4.2.2 To reduce reconfiguration overhead The reconfiguration overhead in dynamically reconfigurable systems is a major problem that affects the total execution time. Resano et al. [154, 155, 156] proposed a specific scheduling algorithm to minimize the reconfiguration overhead by integrating three modules into an existing hybrid run-time/design-time framework called *Task Concurrency Management* (TCM) [221] methodology. The three modules were *reuse*, *prefetch*, and *replacement*. The enhanced version of the TCM scheduler not only reduces power consumption, but also the reconfiguration overhead. Configuration reuse allows different applications to use the same configured hardware task so that the number of configurations are reduced.

Configuration prefetch allows hardware tasks to be configured well before it is required so that the configuration latency is hidden or overlapped with other hardware or software executions. The replacement technique increases the possibilities of reusing those subtasks that are more critical for the system performance.

4.3.4.2.3 To reduce scheduling overhead Two scheduling and placement techniques called *horizon* and *stuffing* were proposed by Steiger at al. [177, 178]. The horizon scheduler maintains an execution list, a reservation list, and a scheduling horizon list. The execution list contains all currently executing tasks t_i with their finishing times $f_i = s_i + e_i$, and placements x_i, where s_i and e_i are, respectively, the start time and the execution time of task t_i, and x_i is a column number. The execution list $\{(t_i, x_i, f_i)\}$ is sorted in increasing order of finishing times. The reservation list stores all scheduled but not yet executing tasks. The reservation list $\{(t_i, x_i, s_i)\}$ is sorted in increasing order of starting times. The scheduling horizon list consists of elements $h_i = ([x_1, x_2]@t_r)$, where $[x_1, x_2]$ denotes an interval in the x-dimension and t_r is the maximum of the last release time for the interval and the current time. The horizon list is stored in increasing order of the release times. When a new task arrives, the horizon scheduler will find a suitable width from the scheduling horizon list for the task. The best-fit strategy is used to find the smallest width in which the task can be placed. Note that in this method, new tasks can only be *appended* to the horizon. It is not possible to schedule tasks before the horizon. The advantage of this technique is that maintaining the scheduling horizon is simple compared to maintaining all future allocations of the device.

Different from the horizon scheduler, the stuffing scheduler employs a free space list instead of the scheduling horizon list. The free space list is a set of intervals $[x_1, x_2]$ that identify currently unused resource intervals, sorted in increasing order of the x-coordinates. When a new task arrives, the stuffing scheduler will find a suitable space from the free areas. The stuffing technique produces better scheduling results than the horizon method, in terms of smaller number of task rejections. However, stuffing is more complex than horizon and thus takes more computation time. Both horizon and stuffing were also extended to the 2D reconfigurable area model [178].

An example comparing the horizon and the stuffing schedulers is illustrated in Figure 4.10, where after placing tasks A, B, C, D, E, F into the reconfigurable area, task G of size 5 columns is to be placed. On the left-hand side of Figure 4.10 is the result of using the horizon scheduling method. The horizon (as represented by the thicker line) allows only tasks of size 4 columns to be placed from time 5 to 19, even though there are actually 5 columns free during that time interval. Task G is thus placed in the area of columns [9,13] in the time interval [20,23]. The stuffing scheduler instead uses the more complex free space list to allow task G to be placed as early as in the time interval

[5,8].

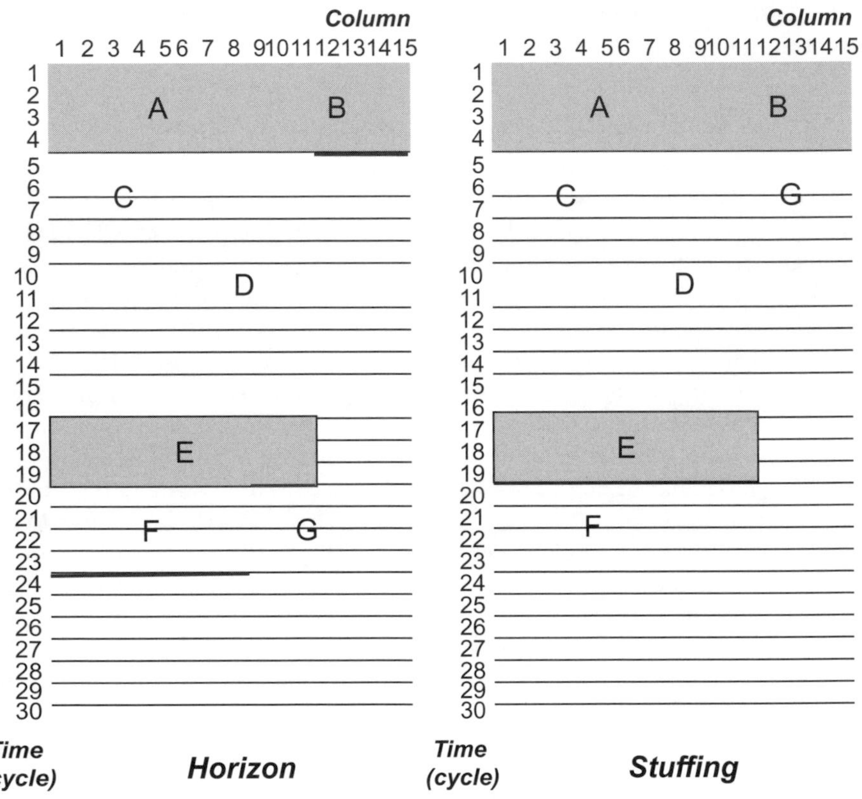

Figure 4.10: Horizon and stuffing schedulers.

4.3.4.2.4 To reduce fragmentation When there is a large variance in the ratios of the size of reconfigurable hardware tasks to their computing times, the fragmentation of allocatable hardware resources becomes a serious problem in both the horizon and the stuffing methods. This is because the different aspect ratios of the tasks produce holes that cannot be used. This is illustrated in left-hand side of Figure 4.11, where task D occupies 4 columns $[7, 10]$ and is scheduled to execute in the time interval $[5, 16]$. There is a big hole in columns [1,6] after task C finishes, that is, $[9, 16]$. This leads to fragmentation and under-utilization of the hardware resources. Hence, a *classified stuffing* technique was proposed by Chen and Hsiung [42] to reduce fragmentation. The main idea is to classify all tasks into two types depending on the *space utilization ratio* (SUR) of a task, where $SUR(t_i) = a_i/e_i$, a_i is the area requirement, and e_i is the execution time of task t_i. If $SUR(t_i) < 1$ then the task t_i is classified as a low SUR task. Otherwise, it is a high SUR task. The classified stuffing method places the low SUR tasks and the high SUR tasks starting from the leftmost and the rightmost column in the reconfigurable area, respectively. The difference in the placement produced by classified stuffing, as compared to conventional stuffing, is illustrated in Figure 4.11, where $SUR(D) < 1$ means task D is a low SUR task and is placed from the rightmost column. The hole is thus not as large as that in the conventional stuffing. The classified stuffing technique shows significant benefits in terms of a shorter system schedule and a more compact placement with smaller fragmentation of allocatable resources.

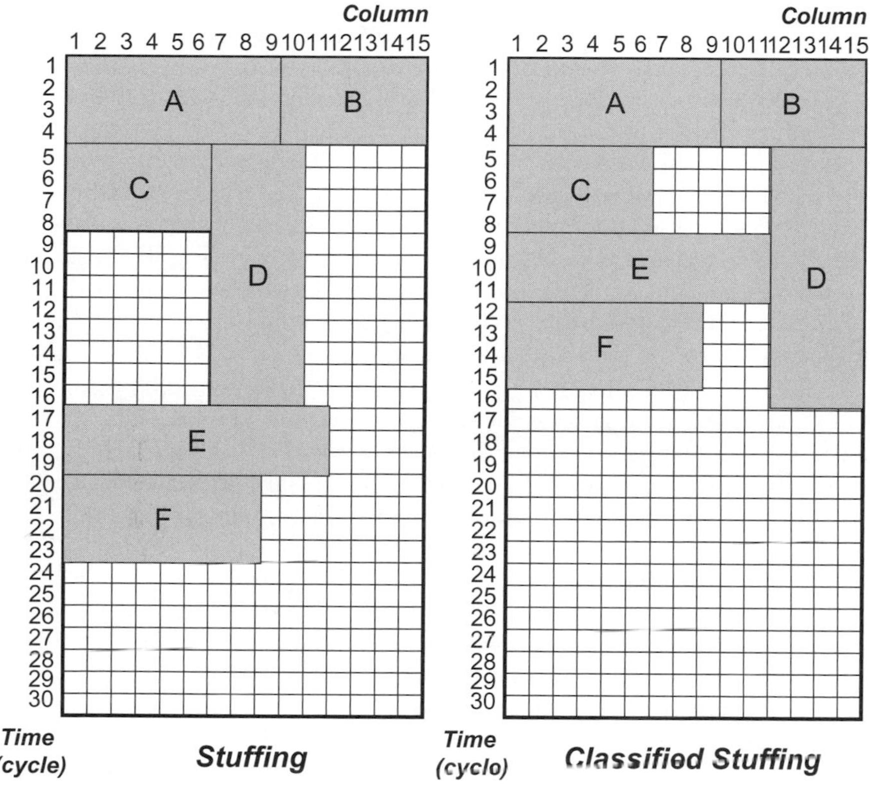

Figure 4.11: Conventional and classified stuffing schedulers.

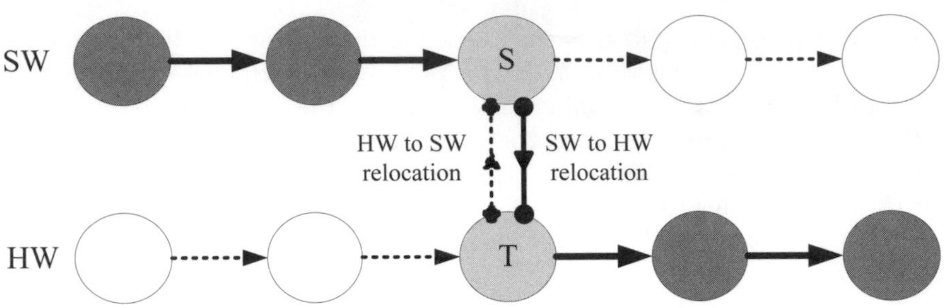

Figure 4.12: Hardware and software task relocations.

4.3.4.2.5 Co-scheduling HW/SW tasks with relocation With proper support, tasks can be migrated from software to hardware and vice versa in dynamically reconfigurable systems. Such tasks are called *relocatable*, which means a software (hardware) task can be preempted and then restarted as a hardware (software) task, as illustrated in Figure 4.12. Such systems require more advanced hardware-software scheduling techniques than the pure hardware schedulers described in the previous sections. The target problem can be stated as follows. Given a dynamically partially reconfigurable system with a reconfigurable resource area of limited size and a set of relocatable tasks with real-time constraints, we need to find a run-time feasible schedule for the task such that all task-related and architecture constraints are satisfied and the hardware resource utilization is maximized.

Pellizzoni and Caccamo [149] proposed a pseudo-optimal allocation algorithm and a relocation scheme for relocatable tasks. The proposed method, which we call *Adaptive Hardware-Software Allocation* (AHSA), consists of a greedy allocation algorithm for generating the initial hardware-software system partition and a group-based, system-wide swapping algorithm for dynamically changing the hardware-software system partition. Software tasks are scheduled using EDF in conjunction with the *Constant Bandwidth Server* (CBS) [1], which provides isolation between hard and soft real-time tasks, so that hard real-time tasks are proven to complete within their deadlines. The AHSA scheduler tries to minimize the total software utilization by maximizing the hardware resource utilization, so that the CPU can allow more tasks to be executed. However, the group swapping in AHSA incurs a quadratic time complexity in terms of the number of tasks. This affects its application to hard real-time systems.

Chiang [43] proposed a method called *Relocatable Hardware Software Scheduling* (RHSS) that takes only linear time complexity because instead of group swapping it tries to minimize the number of reconfigurations by doing incremental configurations into the reconfigurable area. As illustrated in Figure 4.13, given the configurations for a dynamically reconfigurable system and a set of relocatable tasks, RHSS tries to find an initial hardware-software

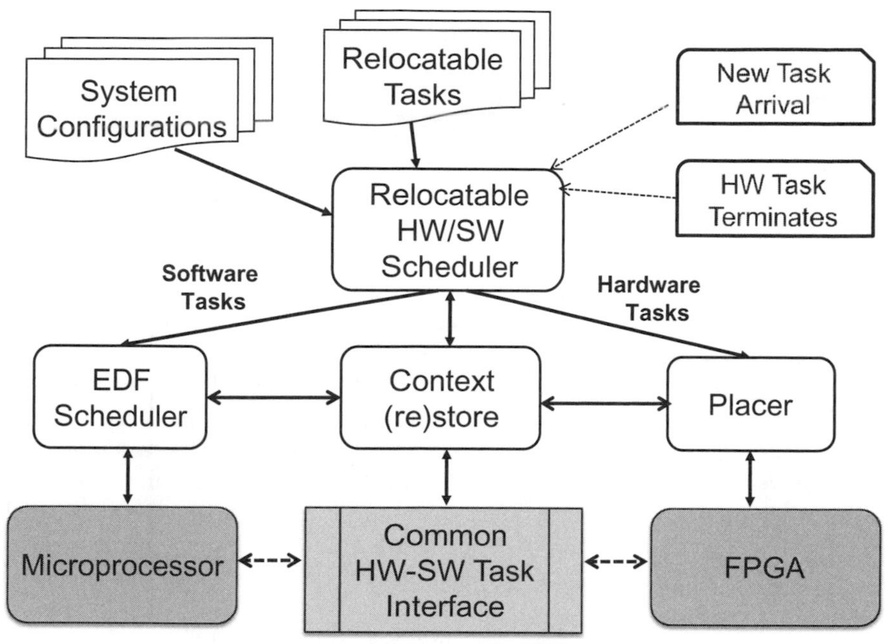

Figure 4.13: Reconfigurable hardware software scheduling.

partitioning such that the software tasks are scheduled by an EDF scheduler on the microprocessor and the hardware tasks are placed on the FPGA for execution. At two different scheduling points, including when a new task arrives and when a hardware task terminates, the RHSS is invoked again. Task relocation further requires the support of two more mechanisms, including context saving and restoring for both hardware and software tasks and a common hardware software task interface such that a relocated task can resume communication with other tasks.

Compared to AHSA, RHSS considers hardware configuration time explicitly and takes contiguous area placement restrictions into account, which results in more realistic models. Application experiments show that in spite of the reduced time complexity, the proposed RHSS method can achieve an even higher resource utilization than AHSA.

4.3.4.2.6 To reduce power consumption Power consumption has always been a major concern in reconfigurable systems because of the increasing static power leakage in deep sub-micron fabrication technology. FPGA may have several parts of the logic powered on, without ever using them, and thus static power leakage becomes a serious issue. Further, the logic configuration in FPGA is through SRAM memory settings that require a large amount of energy. Thus, various methods have been proposed to reduce the power consumption in FPGA-based reconfigurable systems, including both static and dynamic methods. Static methods include optimization of data allocations [161], using Trellis [143] and dynamic priority multirate scheduling [166, 167]. Since OS4RS can only employ dynamic scheduling methods, we do not elaborate on the details of the static methods. Dynamic methods include configuration reuse and replacement strategies, clock gating and frequency-scaling with multi-context and execution scheduling [136, 137, 138], and hardware configuration reuse with processor *dynamic voltage scaling* (DVS) [91, 92, 126].

Noguera and Badia [137] proposed a reconfigurable hardware task scheduling method with power-performance trade-offs. The main techniques used to reduce power consumption were the grouping of execution of tasks to use the same reconfiguration context, and the application of clock-gating and frequency scaling strategies. Two event queues were maintained, one for task execution events and one for task reconfiguration events. The task execution events invoked the execution scheduler and the task reconfiguration events invoked the multi-context scheduler. Compared to a scheduler with a single event queue, the method proposed by Noguera and Badia can achieve an energy reduction of 17.48% at an overhead of 1.96% increase in system execution time.

Well-known low power design techniques for reconfigurable hardware such as those described above and for software such as dynamic voltage scaling (DVS) were individually developed. There was an integration gap between these hardware and software low power design techniques, which may work against each other rather than collaborating towards the overall reduction in system power consumption. For example, if DVS is applied to the execution of a software task, it might delay the execution of a successor hardware task and as a result the hardware task might need an extra configuration because its execution then overlaps another task instance of the same hardware that is already executing.

To fill the integration gap between hardware and software low power design techniques, Hsiung and Liu [91, 92, 126] proposed an energy-efficient hardware-software coscheduling method that integrated hardware configuration reuse with dynamic voltage scaling. The key concept is as follows. Whenever two or more applications have a common hardware task whose executions overlap in time, the coscheduling method tries to arrange the task executions such that the two instances of the common hardware task are sequentially connected in time. For example, if F_i and F_j are two instances of the same hardware task T, then the coscheduling method tries to ensure either $finish(F_i) = start(F_j)$ or $finish(F_j) = start(F_i)$, where $finish()$ and $start()$ are the finish and start times of a task instance.

The coscheduling method takes a set of independent hardware and software tasks as input. A task is represented as 4-tuple $T_i = (A_i, D_i, P_i, FG_i)$, where A_i, D_i, P_i, and FG_i are the arrival time, the deadline, the priority, and the function graph of the task, respectively. A function graph $FG = (V, E)$ is a directed acyclic graph representing the precedence relations among the functions of the task. The goal is to find an execution order for the functions on the processor and one for the FPGA such that all task deadlines, function precedence relations, and FPGA area constraint are satisfied while the schedules consume the least energy.

The coscheduling method [126] consists of two phases, namely a design time phase and a runtime phase. As illustrated in Algorithm 15, at design time, hardware functions that are common to two or more tasks are identified along with their corresponding *leading software functions*, where a leading software function is one that precedes the *common hardware function* in a function graph, with no other software function coming between them along any path. Each path from a leading software function to a common hardware function is called a *key path*. A *delay time-table* is constructed for each pair of key paths corresponding to the same hardware function. The table keeps information about the time difference that is allowed between two leading software functions for the common hardware functions to be scheduled to use the same configuration. The slack time of a leading software function is shared amongst all the corresponding common hardware functions, if there are more than one of them.

Algorithm 15 Energy-Efficient HW/SW Coscheduling: Design Time Phase

Input: A set of tasks $T = \{T_i; i = 1, 2, 3, \ldots, n\}$
Output: A delay-time table
FindCommonHardwareFunctions();
for each common hardware function H **do**
  S = FindLeadingSoftwareFunction(H);
end for
for each key path **do**
  CalculateStartFinishTime();
end for
Δ = ConstructDelayTimeTable();
CalculateSharedSlackTime();

The runtime phase of the energy-efficient hardware-software coscheduling algorithm is illustrated in Algorithm 16. At run time, tasks may arrive arbitrarily at any time instants. Whenever a leading software function is chosen for execution, the scheduler updates and checks the delay time-table to determine if there is another leading software function already executing with a time difference as allowed by the delay time-table. If not found, the leading software function is directly executed, without applying DVS. If found, the common hardware functions are scheduled to use the same configuration and DVS is applied to the currently chosen leading software function. Thus, not only is the number of hardware reconfigurations reduced along with power reduction, but DVS also helps reduce the power usage of software tasks. Hence, the overall effect is the combined efforts of hardware configuration reuse and dynamic voltage scaling.

Algorithm 16 Energy-Efficient HW/SW Coscheduling: Run Time Phase

```
Input: delay-time table
Output: total execution time, total energy consumption, total configuration
energy, % of tasks with deadlines satisfied
for delay-time list in delay-time list table do
  delay-time list = NULL;
end for
while processor ready queue is not empty do
  Choose leading software function S';
  UpdateDelayTimeList();
  if delay-time list associated with S' is empty then
    execute S';
  else
    SlowDown = CheckSlowDown(S');
    if SlowDown = false then
      execute S';
    else
      TuneDownVoltage(S');
    end if
  end if
end while
```

Example 4.1

To illustrate the above described energy-efficient hardware-software co-scheduling phases, we use an example system consisting of 3 tasks, whose task graphs are shown in Figure 4.14. The function associated with a node T_{Xi} is given inside the node as F_j. Functions F_1, F_7, and F_{10} are software, while the rest are all hardware. The software functions are usually the control programs or drivers, while the hardware functions could be any IP such as (I)DCT, matrix multiplier, FFT, motion estimator, that is, circuit designs that could be used across different applications. The arrival times for tasks A, B, C are, respectively, $0, 2600, 5000$, their priorities are $2, 3, 1$, where a higher value represents higher priority, and their deadlines are $6500, 6600, 5500$. The function attributes are as shown in Table 4.1, where SE_i, HE_i, HC_i, and HS_i are the software execution time, the hardware execution time, the hardware configuration time, and the hardware space requirements, respectively, for function F_i.

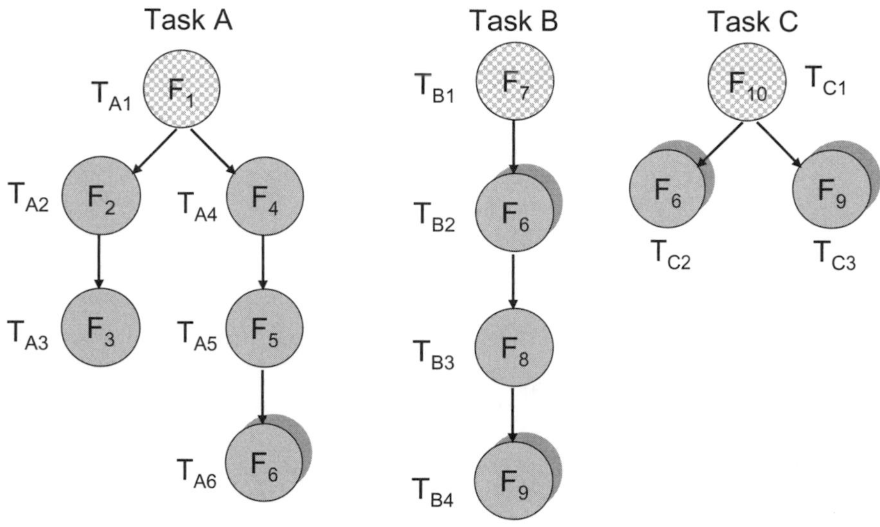

Figure 4.14: An example system with 3 tasks.

In this example, F_6 and F_9 are two common hardware functions, and F_1, F_7, and F_{10} are leading software functions. As shown in Table 4.2, there are 5 key paths from the 3 software functions to the common hardware functions. On applying the design time phase of our proposed scheduling algorithm, the delay time table constructed for this example is as tabulated in Table 4.2. Since there is only one leading software function in each task, the shared slack time is equal to the task slack time, i.e., $300, 100, 1900$ for tasks A, B, C, respectively.

Using the delay time table in Table 4.2, the run time phase is applied as follows. For the leading software function F_1, there is no active key path at time 0, so its execution is not prolonged, and it is executed at full speed of the processor.

At time 2600, task B arrives, and leading software function F_7 is selected for scheduling. There is one active key path, that is $P_1 = \langle F_1, F_4, F_5, F_6 \rangle$, which has hardware function F_6 common with path $P_2 = \langle F_7, F_6 \rangle$. Since the difference between completion times of the leading software functions, $6100 - 2600 = 3500$ is within the time interval $\Delta(1,2) = [1200, 3600]$, so the execution time of F_7 can be prolonged by $3600 - 3500 = 100$ time units, that is, extended from 3500 to 3600. Table 4.3 shows the 10 different voltage levels and the corresponding frequencies and power consumption for the microprocessor used in this example. The reconfiguration power for a CLB column is 450 mW

Table 4.1: Function Attributes for Example System

F_i	SE_i (μs)	HE_i (μs)	HC_i (μs)	HS_i
F_1	2600	–	–	–
F_2	–	700	600	1
F_3	–	1000	600	1
F_4	–	1200	600	1
F_5	–	1200	600	1
F_6	–	1200	1200	2
F_7	3500	–	–	–
F_8	–	1200	600	1
F_9	–	600	600	1
F_{10}	2400	–	–	–

Table 4.2: Delay Time Table for Example

key path	P_1	P_2	P_3	P_4	P_5
P_1	X	[1200,3600]	X	[0,1200]	X
P_2	X	X	X	[0,1200]	X
P_3	X	X	X	X	[1800,3000]
P_4	X	[0,1200]	X	X	X
P_5	X	X	X	X	X

P_1: $\langle F_1, F_4, F_5, F_6\rangle$, P_2: $\langle F_7, F_6\rangle$, P_3: $\langle F_7, F_6, F_8, F_9\rangle$, P_4: $\langle F_{10}, F_6\rangle$, P_5: $\langle F_{10}, F_9\rangle$, X: undefined

and the execution power in FPGA is 1000 mW. The prolongment of F_7 results not only in lesser energy consumption by the software function F_7, but also saves one reconfiguration of two columns of FPGA due to the configuration reuse now possible for hardware function F_6 common to tasks A and B, along key paths P_1 and P_2, respectively. Fewer reconfigurations save both time and energy.

At time 5000, task C arrives, but has to wait till time 6200 for the processor to be available to execute leading software function F_{10}. Figure 4.15 shows the details of why and how the execution of F_{10} can be prolonged by 600 time units by applying our co-scheduling method. Further, one more reconfiguration of 1 column FPGA is saved for F_9 common to tasks B and C, along paths P_3 and P_5, respectively. The final scheduling results are shown in Figure 4.15.

Let us compare the energy-efficient hardware-software coscheduling method with two conventional methods as follows. Method M_1 does not apply any acceleration technique, while method M_2 applies configuration prefetch and reuse, but without DVS. The comparisons are shown in Table 4.4, where we can see that the energy-efficient coscheduling method outperforms the other two methods. Compared to the bare method M_1, the energy-efficient

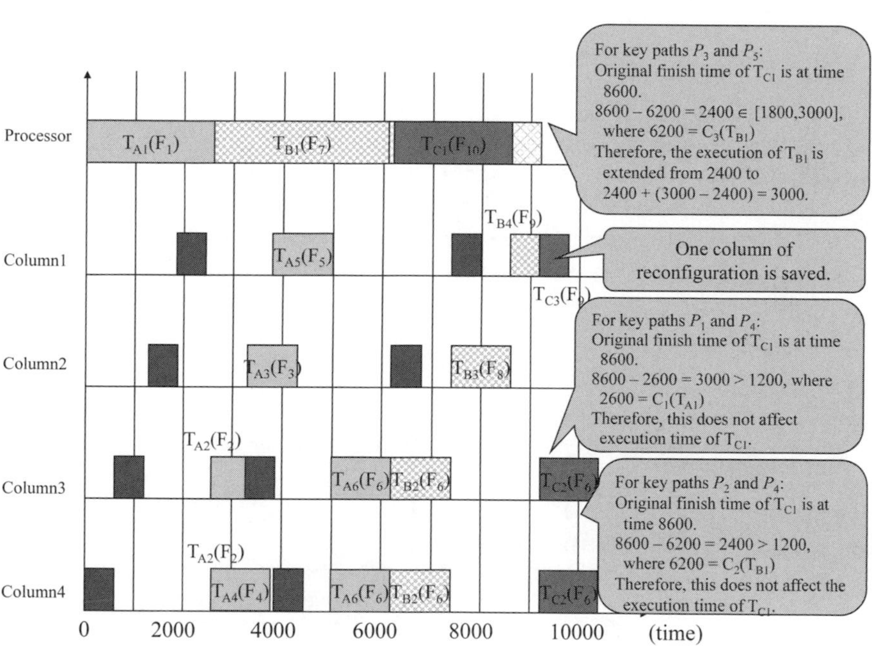

Figure 4.15: Scheduling results for example.

Table 4.3: Processor Voltages, Frequencies, Power Consumption

	Voltage (V)	Frequency (MHz)	Power (W)
1	1.750	1000	9.50
2	1.670	944	6.92
3	1.600	912	5.30
4	1.500	868	4.20
5	1.350	812	3.00
6	1.225	776	1.86
7	1.200	709	1.34
8	1.150	655	1.10
9	1.100	590	0.75
10	1.000	545	0.58

coscheduling method not only consumes lesser amounts of total energy and total configuration energy by 24.23% and 38.5%, respectively, and allows 33.3% more task deadlines to be satisfied, but the total execution time is also reduced by 16%. Compared to method M_2, the energy-efficient coscheduling method similarly satisfies all task deadlines, but it requires an additional execution time of about 7.2%, while saving 24.1% total energy and 33.3% configuration energy.

Table 4.4: Scheduling Results Comparison

	M_1	M_2	M	$I_1(\%)$	$I_2(\%)$
TT	12100	9700	10400	−14.0	+7.2
TE	96.76	96.49	73.27	−24.3	−24.1
CE	3.51	3.24	2.16	−38.5	−33.3
RN	2/3	1	1	−33.3	0.0

TT: total execution time (μs), *TE*: total energy consumption (mJ)

CE: total configuration energy (mJ),

RN: % of tasks with deadlines satisfied, $I_i = (M - M_i)/M_i$

□

4.3.5 Module Layer

We talked about hardware task configuration reuse and prefetch in both the placement and the scheduling layers. However, the ultimate decisions on which application uses which task configuration and how the reuse and prefetch techniques are to be designed are the features to be supported in the *module layer*. A good example of this implementation is the *Core Management* proposed by Beretta [21].

A hardware task function can be physically represented by a bitstream such as that for Xilinx FPGAs. All hardware task functions must be registered with the module layer to be managed. A registered hardware task function is called a *module*. The main functions of a module layer are the selection of modules and the efficient management of modules in the system. Take the discrete cosine transform (DCT) as an example module, which can be used across different multimedia applications such as MPEG4 or H.263. Different implementations of the same function, such as 1D or 2D transforms, which have different performance and resource requirements could be available for selection in the module layer. Hence, based on the current reconfigurable hardware resource usage, the module layer can decide which module implementation to allocate to a particular application request.

As illustrated in Figure 4.16, after a module is instantiated, it could be in one of several states, including *instantiated*, *configured*, *cached*, *computing*, *waiting*, *preempted*, and *terminated*. The details of each state are described as follows.

- *Instantiated*: A module is requested by an application and an instance of the requested module is created. However, it must wait for its turn to be scheduled and placed.

- *Configured*: A module is already scheduled, placed, and configured in this state. However, it is not yet running. This is more like the ready state in conventional processes.

- *Cached*: A module is already *released* by an application, but its resources are not de-allocated. The module is still configured, but unassigned to any application. A cached module can be reused, without configuration.

- *Computing*: An application has already started using a configured module. A computing module may communicate with other modules if a communication infrastructure exists.

- *Waiting*: A module is waiting for some data from an application or for some generic events such as I/O, etc.

- *Preempted*: If supported by the reconfigurable system architecture, a module could be preempted while computing. In this state, the context of the module is stored in some memory so that it can be restored later in another hardware module instance or in some relocated software task.

- *Terminated*: A module is terminated and its termination status will be retrieved by the operating system.

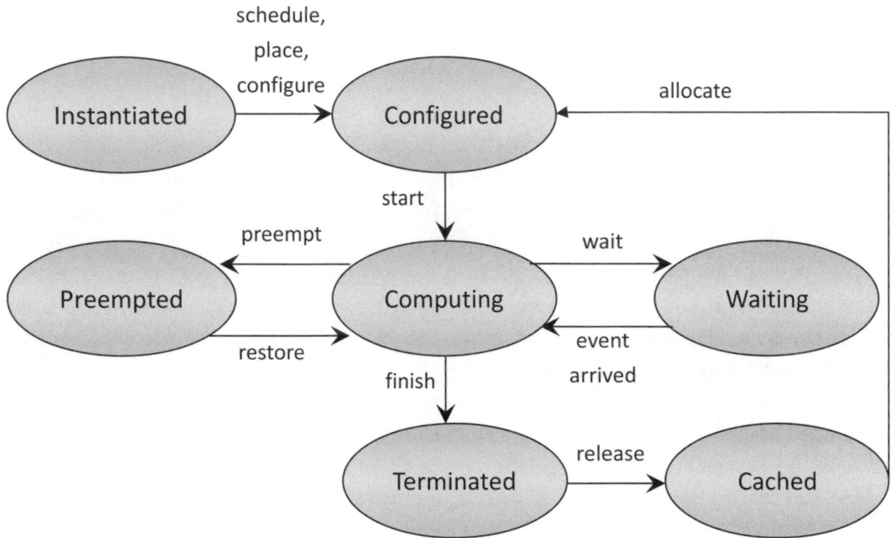

Figure 4.16: State diagram of a generic module.

Whenever an application makes a request for a module, the module layer selects either an existing unused module that is in the *cached state* or instantiates a new one for scheduling, placement, and configuration by the underlying layers. Once a module is configured, it enters the *configured state*, waiting for further invocation by the requesting application. With the *start* signal, the module begins execution by entering the *computing state*. During computation, the module could wait for some events such as application data or signals, or it could be preempted and then restored later, as either a software or hardware task.

For efficient management of modules, they can be *cached*. After a module execution terminates and is released by an application, it is not de-allocated from the reconfigurable logic resource area. The allocated resources are still occupied by that module. This is the *cached* state of a module. If there are enough resources or few requirements for modules, caching a module saves both configuration time and power. Thus, caching is one of the most important functions of the module layer.

The main functions of the module layer include the following, all of which can be implemented as system calls in the OS4RS.

- *Module Request*: an application can make a request for a module of some identification label M_i. Upon receiving a module request, the OS4RS selects either an existing cached module to satisfy this request or instantiates a new module. Then, the OS4RS performs the necessary scheduling, placement, and configuration by retrieving the bitstream from some external memory.

- *Module Reset*: an application can reset an allocated module based on its own requirements and decisions.
- *Module Start*: an application can invoke an allocated module to start computing.
- *Module Release*: an application can release an allocated module when it does not require the module anymore. However, the module layer does not reset the resources occupied by the released module; instead, the module is cached for possible future use.
- *Module Delete*: an application can request a module to be completely eliminated from the system, that is, the module is not only released, but its allocated resources are also released.
- *Module Shared*: an application sets an allocated module to be shareable with other applications. However, the synchronization between applications sharing the same module is not the responsibility of the module layer. It is left to the applications designers.
- *Module Unshared*: an application sets an allocated module to be unshareable. This is the default mode for a newly allocated module.
- *Module Preempt*: an application requests a module to be explicitly preempted, that is, its context data and state are stored and the module is swapped out from the system, which means the module could be either cached or deleted, depending on the implementation of the OS4RS.
- *Module Restore*: an application requests a module to be explicitly restored, that is, its stored context data and state are retrieved and restored. This could amount to a new scheduling, placement, and configuration if the module is no longer cached.
- *Module Cache Clear*: an application could request all cached modules to be cleared. This helps in generating the largest amount of available free resources for an application.

It is not necessary for an OS4RS to design and implement a complete module layer. However, a simple one is required for efficient management of modules. There are several trade-offs in the design of such a layer. For example, different caching policies could be used [21] to make the system performance more efficient. Different module sharing schemes could lead to different system power usage and efficiency.

4.3.6 Application Layer

The topmost layer of an OS4RS is the application layer. An application is defined as a set of communicating tasks, which could be implemented as either software or hardware or both. All software tasks are loaded, scheduled, and executed by the OS4RS. All hardware tasks are also scheduled, placed, and configured for execution by the OS4RS. For efficient communication, the OS4RS must also provide a uniform communication infrastructure between the software and hardware tasks.

Some typical examples of application include an image compression application such as a JPEG encoder, which consists of three tasks such as DCT, Quantization, and Huffman encoding. DCT could be implemented as a hardware task module and then shared with other image compression applications.

In general, an application consists of the following information.

- *Application Name*: This is the name for an application and it can be used to identify the application. For example, JPEG is an application name.
- *Task Graphs*: This is a set of graphs representing the functions of an application. Each task graph $G = (V, E)$ consists of vertices in V that represent the required tasks and edges in E that represent the precedence relationships among the tasks. The precedence relationship could also mean the requirement for a communication channel for data transfer.
- *Arrival Time*: This is the time when an application is triggered.
- *Period*: The period for a periodic application such as an infra-red sensor in a thief-detector machine.
- *Deadline*: The deadline for completing one job of an application. This is required for real-time applications such as the braking system in a car. It is optional for other applications.
- *Non-functional Constraints*: An application may be constrained to consume at most a certain power limit or memory limit. Such non-functional constraints may be specified for an application.

Each task is further defined as a basic unit of computation and is represented as $t_i = (a_i, p_i, d_i, he_i, se_i, hc_i, hs_i)$, where the attributes are as follows.

- *Arrival Time* (a_i): This is the time when a task is triggered for execution.
- *Period* (p_i): This is the task period if the task is a periodic one.
- *Deadline* (d_i): This is the task deadline if the task is a real-time task.

- *Hardware Execution Time* (he_i): This is the hardware execution time.
- *Software Execution Time* (se_i): This is the software execution time.
- *Hardware Configuration Time* (hc_i): This is the hardware configuration time.
- *Hardware Configuration Space* (hs_i): This is the hardware configuration space requirements.

In the JPEG encoder, the DCT, Quantization, and Huffman encoding can be represented by the attributes listed in Table 4.5.

Table 4.5: JPEG Task Attributes

Task	a_i	p_i	d_i	he_i	se_i	hc_i	hs_i
DCT	0	1	20	100	500	50	5
Q	0	2	10	50	200	40	4
HE	0	3	10	40	200	40	4

Q: Quantization, HE: Huffman Encoding
a_i, p_i, d_i, he_i: IP cycles, se_i: processor cycles,
hc_i: configuration cycles, hs_i: columns

4.4 OS4RS Examples

Different OS4RS were proposed and implemented for reconfigurable systems. Table 4.6 summarizes the OS4RS and compares them. The features that are compared include FPGA platform, reconfiguration mode, partial reconfiguration capability, relocation support, and preemption support.

As we can observe from Table 4.6, MARC.1 was the first OS4RS that was developed in 2001. There have been different operating systems proposed throughout these years. Mostly the target platform was Xilinx Virtex II series FPGA chips, except for the OO/NoC simulation done in SystemC and the BORPH OS on the BEE2 platform. The reconfiguration mode is either internal or external. Internal reconfiguration is achieved by using a reconfiguration controller such as ICAP and its controller and driver. External reconfiguration is mostly done using the JTAG port. Support for partial reconfiguration is a basic requirement for modern OS4RS. BORPH was originally targeted at multi-FPGA systems, hence partial reconfiguration support was not designed. Support for relocation requires the design of either a software or a

Table 4.6: Comparing Different OS4RS

OS4RS	Refs	Year	FPGA platform	RM	PR	RS	PS
MARC.1	[204, 206]	2001	XV	Ext	Yes	No	No
ReConfigME	[208, 207]	2002	XV	Ext	No	Yes	No
OS4RS	[200]	2003	XCV-800	Int	Yes	No	No
Egret/μCLinux	[209]	2004	XV II	Int	Yes	No	No
IPCM	[58]	2005	XV II-Pro	Int	Yes	No	No
OO/NoC	[83]	2006	SystemC	Ext	Yes	Yes	No
BORPH	[176, 175]	2006	BEE2	Ext	No	No	No
CoreMgt OS	[21]	2008	XV II-Pro	Int	Yes	Yes	Yes
CCU OS4RS	[93]	2008	XV 4/5	Int	Yes	No	Yes

RM: Reconfiguration Mode, PR: Partial Reconfiguration,
RS: Relocation Support, PS: Preemption Support
XV: Xilinx Virtex, Ext: External, Int: Internal

hardware filter for dynamic modification of hardware bitstreams. Out of the nine OS4RS listed in Table 4.6, only three OS4RS have the relocation support and two OS4RS have the preemption support. From this observation, we can conclude that these advanced features can be found only in the more modern OS4RS such as Core Management [21] and the CCU OS4RS [93].

Summary

We introduced the motivation and the requirements for the design of an operating system for reconfigurable systems. A layered architecture for OS4RS was also described in details. The design of several algorithms in an OS4RS was also described. Finally, some real-world examples of OS4RS were cited and compared. Though it is yet not possible to have a full-fledged OS that meets all the requirements of an OS4RS that we have described, we believe we will soon see such an OS as long as there is a consensus on the architecture of reconfigurable systems. Interested readers may further explore the extensive literature on this topic, some of which is given in the references.

Exercises and Problems

1. Why will a traditional operating system not suffice for managing the tasks in a reconfigurable system? Give all the possible reasons that you can think of.

2. Which functions of an OS4RS do you think are the most difficult to implement? Why?

3. What do you think are the advantages and disadvantages of the layered architecture for an OS4RS?

4. Which layers in the architecture for an OS4RS are the minimally required?

5. Which scheduling algorithms do you think are feasible for run-time usage?

6. What are the constraints of the current FPGA technology, including both device constraints and tool constraints, that affect the development of a full-fledged scheduler and placer in a reconfigurable system?

7. For dynamic partitioning to be feasible, what kind of support must the system architecture provide?

8. **Project Theme**: Try to design a hardware task scheduler and placer at the user level in an existing operating system such as the latest version of Linux OS. You can either refer to the existing literature on the scheduling and placement algorithms or try to develop your own. Note that you need to first introduce the concept of a hardware task into the OS. You also need a well-defined reconfigurable system architecture to support your scheduling and placement.

5

Dynamic Reconfigurable Systems Design Flows

After presenting different solutions proposed to design and implement dynamic reconfigurable systems, this chapter will describe a general and complete design methodology that can be followed as a guideline for designing reconfigurable embedded systems. The proposed workflow aims at designing a complete framework able to support different devices (i.e., [102, 101]), multi-FPGA architecture (i.e., the RAPTOR2000 system [109]), different reconfiguration techniques ([95, 98]) and type of reconfiguration (i.e., internal or external, mono and bi-dimensional) that allow a simple implementation of an FPGA system specification, exploiting the capabilities of partial dynamic reconfiguration. The idea behind the proposed methodology is based on the assumption that it is desirable to implement a flow that can output a set of configuration bitstreams used to configure and, if necessary, partially reconfigure a standard FPGA to realize the desired system. One of the main strengths of the proposed methodology is its low-level architectural independence. This chapter can be seen as a bridge between Chapters 3 and 4 and Chapter 7.

5.1 System Design Flows

To develop a configurable or a reconfigurable system it is possible to build an ad hoc solution or to follow a generalized design flow. The first choice implies a considerable investment in terms of both time and effort to build a specific and optimized solution for the given problem, while the second one allows the re-use of knowledge, cores, and software to reach a good solution to the same problem more rapidly. Section 5.1.1 describes the former approach, presenting an overview of the state of art of all the design techniques that can be considered as ad hoc solutions or as basic flow, while Section 5.1.2 proposes an overview on the complete design flows that have been implemented over the last years to design reconfigurable systems.

5.1.1 Basic Flows

From a general point of view, as described in [95], partial reconfiguration can be performed following two different approaches: **module-based** or **difference-based**. The first one is the *module-based* approach, which is characterized by the division of reprogrammable devices in a certain number of portions, each one of which is called reconfigurable slot. In this scenario it is possible to reconfigure one or more reconfigurable slots with a hardware component that is able to perform a specific functionality, called module. Obviously, the modules contained in slots that are not involved in the reconfiguration task do not have to stop during the reconfiguration process. The *difference-based* approach does not require slots and modules definition, but it is only suitable when the differences between two configurations are very small since the process on which it is based is suitable only when small changes in the design are required. The most general design approach for dynamically reconfigurable embedded systems, as described in [95], is the modular-based design. This approach is strongly connected to the module-based reconfiguration approach and is based on the idea of a design implemented considering the system specification as composed of a set of several independent modules (called IP-Cores, Intellectual Property-Cores) that can be individually synthesized and finally assembled to produce the desired system.

Nowadays, a novel design flow based on the module-based approach has been introduced: the Xilinx **Early Access Partial Reconfiguration** [98] (EAPR). It is a Xilinx design flow,* based on *Integrated Synthesis Environment* (ISE), extended with the proper EAPR patch. This approach extends the previous one, introducing three new features:

- The first one is the possibility to allow signals (routes) in the base design to cross through a partially reconfigurable region without the use of a bus macro.
- The second interesting characteristic of the EAPR flow is the possibility of defining rectangular reconfigurable regions (2D placement), since there are no more constraints that make it necessary to use the whole height of the reprogrammable device for each reconfigurable region. Unfortunately, this feature concerns just the constraints definition, since it is not possible† to define two reconfigurable regions that share two different parts of the same column, so it is compulsory to assign a column to a single reconfigurable region. This observation explain why, even using a 2D placement, it is not possible to perform a 2D reconfiguration.
- The last feature is that this new approach supports Virtex-4 [101] and [100] devices.

*Developed exclusively for the 8.2, 9.1, and the 9.2 versions of ISE.

†Using Virtex, VirtexII and VirtexIIPro Xilinx devices [218].

In such a scenario it can be noticed how this new flow provides a deep improvement in the design of reconfigurable architectures but without changing the context in which the previous one also was working. Considering some aspects of the reconfiguration process, i.e., the bitstream reallocation scenario; it can be easily described via the difference bitstreams [85, 65, 119, 47, 118, 108] and via *blank* modules approaches and only with the introduction of a set complex constraints that will lead to the loss of all the presented advantages in the *EAPR* case.

5.1.2 Generic Flows

The aim of the **ADRIATIC** project [148] is to define a methodology able to guide the codesign of reconfigurable SoC, with particular attention to cores belonging to the wireless communication application domain. The first phase is the system specification, in which the functionality of the system can be described by using a high-level language program, like in a standard design flow. This executable specification can be used to accomplish the following tasks:

- generation of the test-bench, which can be used in the other phases of the design,
- partitioning of the application to specify which part of the system will be implemented in hardware (either static or dynamically reconfigurable hardware),
- accurate definition of the application domain and the designer knowledge.

To derive the final architecture from the input specification, the dynamically reconfigurable hardware has to be identified; each dynamically reconfigurable hardware block can be considered as a hardware block that can be scheduled for a certain time interval. During the partitioning phase it has to be decided for each part of the system, if it has to be implemented in software, in hardware, or in a reconfigurable hardware block. To aid in this decision, some general guidelines have been developed. In the mapping phase the functionalities defined by the executable specification are modified to obtain thorough simulation results. It is possible to state that the *ADRIATIC* flow is a solution that can easily be applied to the system level of a design. In this phase, in fact, it is possible to draw benefits from the general rules that guide the partitioning and from the mapping phase. However, there is no detailed description of the following phases, which take place at RTL level, thus there are some implementation problems that cannot find a solution within the *ADRIATIC* flow.

The **RECONF2** [30] aim is to allow implementation of adaptive system architectures by developing a complete design environment to benefit from dynamic reconfigurable FPGAs; in particular it is targeted to real-time image processing or signal processing applications. The *RECONF2* builds a set of partial bitstreams representing different features and then uses this collection to partially reconfigure the FPGA when needed; the reconfiguration task can be under the control of the FPGA itself or through the use of an external controller. A set of tools and associated methodologies have been developed to accomplish the following tasks:

- automatic or manual partitioning of a conventional design
- specification of the dynamic constraints
- verification of the dynamic implementation through dynamic simulations in all steps of the design flow
- automatic generation of the configuration controller core for VHDL or C implementation
- dynamic floorplanning management and guidelines for modular back-end implementation

It is possible to use as input for this flow a conventional VHDL static description of the application or multiple descriptions of a given VHDL entity, to enable dynamic switching between two architectures sharing the same interfaces and area on the FPGA. The steps that characterize this approach are the partitioning of the design code, the verification of the dynamic behavior, and the generation of the configuration controller. The main limitation of the *RECONF2* solution is that there is not the possibility to integrate the system with both a hardware and a software part since both the partitioned application and the reconfiguration controller are implemented in hardware in the final system.

The works proposed in [202, 25, 39] are all examples of modular approaches to the reconfiguration based on the Xilinx module-based [95] technique. The flow proposed in [25] is too *human based*, it is basically a proof of concepts of the Xilinx methodology demanding a deep interaction from the designer, who has also to be a very highly skilled designer in matter of partial reconfigurable design techniques, in all the phases of the flow. In [202] the reconfigurable computing platform is intended to be PC based. In such a context the host PC will take care of the download of all the necessary files to partially reconfigure the underlying reconfigurable architecture. Such architecture has been defined using a Xilinx FPGA, which can be accessed using common PC BUS, i.e., PCI, USB, FireWire. The **Proteus** framework [202] does not provide substantial improvements in the definition of a novel design flow for partial reconfigurable architectures. This framework can be useful to develop applications that will find benefits in the execution over the proposed reconfigurable platform, since Proteus will take care of the communication between the host PC and the reconfigurable architecture, it will manage the download of the necessary partial reconfiguration bitstreams, but it presents all the drawbacks already indicated in the Xilinx standard modular-based design flow. The **PaDReH** framework, described in [39], aims at designing dynamically and partially reconfigurable systems based on single FPGAs. The flow is composed of three phases: *Design Capture and Functional Validation*, *Partitioning and Scheduling*, and *Physical Synthesis and Reconfiguration Infrastructure*. Even if all these three phases are mentioned it is possible to read that just the last one has been implemented. The main contribution of this work can be found in the definition of a hardware core used to manage the reconfiguration since the bitstream generation phase is basically the same proposed by Xilinx in [95].

In [165] two different techniques for implementing modular partial reconfiguration using Xilinx Virtex FPGAs are proposed and compared. The first method has been already introduced and widely commented. It is the design flow, proposed by Lim and Peattie in the Xilinx application notes [95]. The second technique, proposed by the authors in [165], has removed the monodimensional constraints and it can be considered as the ancestor of the EAPR [98] approach. This method, called **Merge Dynamic Reconfiguration**, is based on a *bitstream merging process* and *reserved routing*. Due to the reserved routing, it is possible to have statically routed signal to pass through a reconfigurable area. Since the static and the reconfigurable modules are placed and routed independently it is possible to reserve already used routing resources from the static core to prevent the reconfigurable module from using the same resources. In this context the separation between the design of the static component and the reconfigurable ones is clearly stated, but the drawback is the reduction in the freedom of the router. If a partial reconfiguration bitstream has to be downloaded on the FPGA, the Merge Dynamic Reconfiguration approach does not write it directly to the reconfiguration memory but instead reads back the current configuration from the device and updates it with the data from the partial bitstream in a frame-by-frame manner, minimizing the amount of memory requested to store the bitstream. The result of this method is the possibility of having two or more modules overlapping their vertical placement constraints, allowing modules to be shaped and positioned arbitrarily, but it has the drawback of dramatically incrementing the reconfiguration time.

5.2 Reconfigurable System Design Flow: Structure and Implementation

The idea behind the methodology presented in this chapter is based on the assumption that it is desirable to implement a flow that can output a set of configuration bitstreams used to configure and, if necessary, partially reconfigure a standard FPGA to realize the desired system. One of the main strengths of the proposed methodology is its low-level architectural independence. The software side of the desired solution can be developed either as a standalone application or with the support of an operating system, such as Linux. Even if the standalone approach can be optimized for each particular scenario to improve timing performance, it reduces the flexibility of the whole system, so it is usually desirable to employ a standard operating system. See Chapter 4 for more information.

The aim of the flow presented in this chapter[‡] is to generate the implementation of the desired system in order to make it possible to physically configure the target device to realize the original specification. A diagram showing the whole process is presented in Figure 5.1.

To develop the system it is necessary to consider the design flow as composed of three parallel parts: the hardware (HW), the reconfigurable (RHW), and the software (SW) sides.

On one hand the first steps that have to be performed in the hardware and reconfigurable hardware sides are the Hardware Description Language (HDL) *Core Design* and the *IP-Core Generation*, in which the core functionalities of the original specification are translated in a hardware description language and extended with a communication infrastructure that makes it possible to interface them with a bus-based communication channel. After these steps, the static components of the system are used to realize the design of the static architecture, while the reconfigurable components are handled in a different way, as reconfigurable IP-Cores (RFU); in other words, they will be kept separated from the static part of the architecture, while the static components will be synthesized together with the static part of the architecture.

In order to implement the correct communication infrastructure between the static component and the reconfigurable ones, the component placement constraints need to be known. Therefore, before the generation of the HDL description of the overall architecture, the placement constraints phase needs to be executed. We decided to consider the system description and the design synthesis and placement constraints assignment phases as separate even if, as shown in Figure 5.1, they are extremely interrelated. In fact, the system description phase is needed to generate the HDL architectural solution while the

[‡]Based on the work proposed in [32, 152, 163].

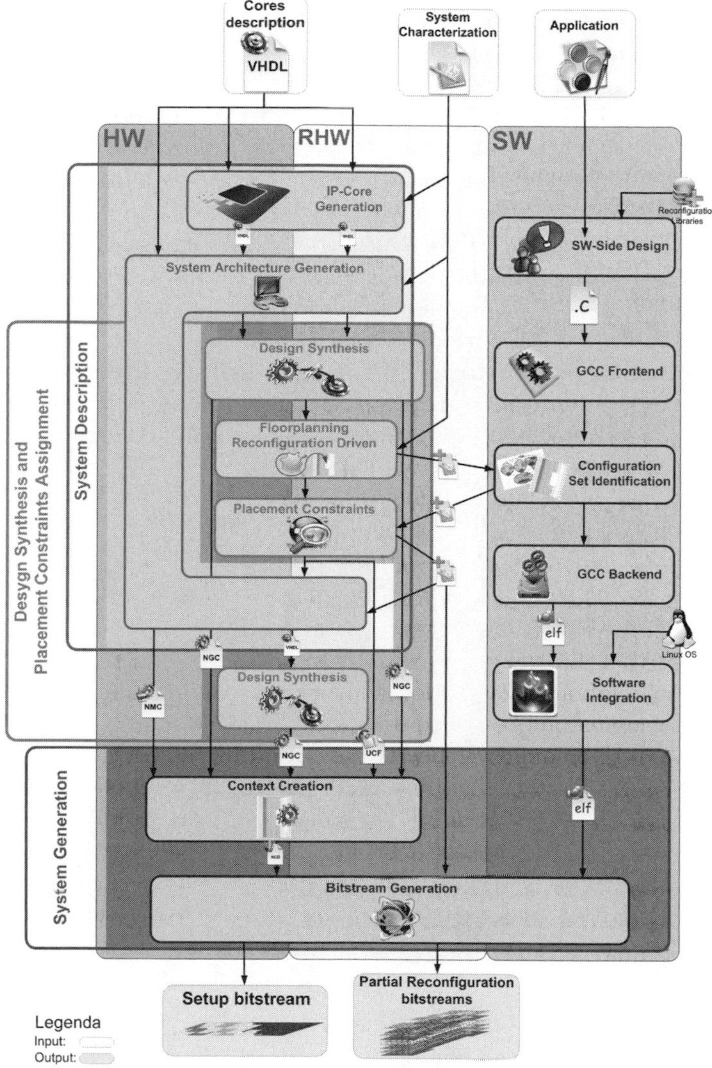

Figure 5.1: The overall view of the described design flow.

design synthesis and placement constraints assignment find its rationale in the identification and in the corresponding assignment of the physical placement constraints needed to define the overall structure of the final architectural solution.

On the other hand, in the reconfiguration management software part, there is the need to develop, in addition to a control application that is able to manage the reconfiguration tasks, a set of drivers to handle both the reconfigurable and the static components of the system. All these software applications are compiled for the processor of the target system. The compiled software is then integrated, in the *Software Integration* phase, with the bootloader, with the Linux OS and with the Linux Reconfiguration Support, which extends a standard OS, making it able both to perform reconfigurations of the reprogrammable device and to manage the reconfigurable hardware as well as the static hardware, in order to allow component runtime plugin. The following step is the *Bitstreams Generation*, which is necessary to obtain the bitstreams that will be used to configure and to partially reconfigure the reprogrammable device. Finally, the last step of the design flow is the *Deployment Design* phase, which aims at creating the final solution that consists of the initial configuration bitstream, the partial bitstreams, the software part (bootloader, OS, Reconfiguration Support, drivers, and controller) and the deployment information that will be used to physically configure the target device.

The self-dynamic reconfigurable architecture that will be used has target architecture that is defined by a static and a reconfigurable side. The physical implementation of such an architecture, organized in three different layers, is shown in Figure 5.2.

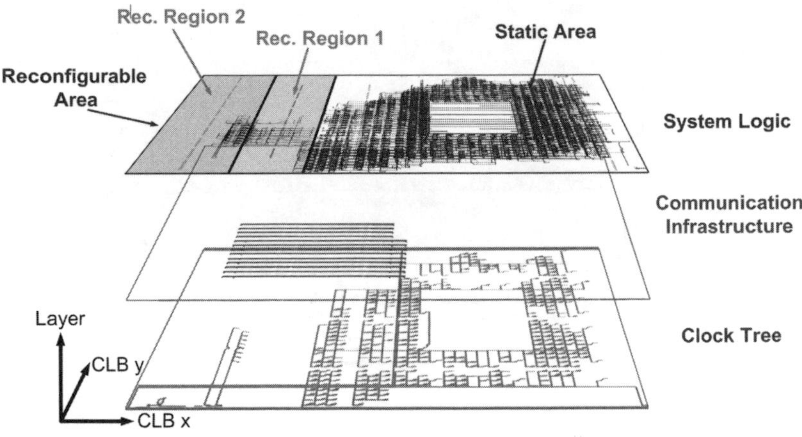

Figure 5.2: Overview of the target self-reconfigurable architecture.

Starting from the bottom, Figure 5.2 shows the reference reconfigurable architecture implemented over a Virtex II Pro VP7 (the layer containing the BRAM memories and Power-PC 405 has not been depicted). The first layer is the clock level that is, according to Xilinx Documentation [216, 95, 218], routed at a different level with respect to the other signals. The second layer contains the communication infrastructure layer between the static and the reconfigurable side. The last one contains CLBs and Switch Matrices and consequently all user logic, and the CoreConnect components are also implemented at this level. The architecture foresees a static and a reconfigurable portion; the static one is typically designed using standard tools, such as EDK [99], or manually described in VHDL. It is composed of a processor (that can be hardcore, i.e., a PowerPC, or softcore, i.e., a Xilinx Microblaze [215]) and a set of cores used to implement an appropriate bridge to the reconfigurable portion. It is important to notice how this kind of self reconfigurable architecture can be implemented using both the PowerPC and the Microblaze because of the difference in the Xilinx FPGAs, i.e., a PowerPC architecture can be used with VirtexIIPro FPGAs, but it is not available in the majority of the Virtex4 and Virtex5 devices. The IBM CoreConnect [48] has been chosen as the main communication infrastructure, and both its OPB and PLB buses will be used to interface the hardware relocation filter proposed in this work. Whenever a PowerPC-based architecture will be used the PLB version of the core will be plugged into the system, while in a Microblaze solution the OPB version of the core will be used. The processor will be used to execute the software side of the application and to correctly manage the necessary calls to the hardware components (with the corresponding reconfiguration, if necessary) designed to speed up the overall computation. The reconfigurable part of the architecture is a portion of the FPGA area on which different reconfigurable functional unit can be mapped, with only the requirement to implement an appropriate bus interface. All the necessary configuration bitstreams can be either stored in an internal memory or eventually loaded by external devices. In any case, these files have to be be sent on the bus to be used by the reconfiguration controller. In this way an overhead is created on the bus that can become slower than the component that manages the reconfiguration process. This scenario can bring to a slowdown of the entire process. Thus it is necessary introduce some sort of mechanism that can manage to accelerate it.

5.3 The Hardware Side of the Design Flow

5.3.1 System Description Phase

The first phase of the design flow, as shown in Figure 5.3, consists of the creation of all the necessary files used to describe the reconfigurable solution, i.e., the VHDL descriptions of each reconfigurable component, the VHDL description of the overall architecture, and the macro HW used to define the communication infrastructure. This phase accepts as input the VHDL descriptions of the core logic used to implement the desired application and the system description file containing the information regarding the overall solution, i.e., the number of reconfigurable slots, the need of the runtime relocation support.

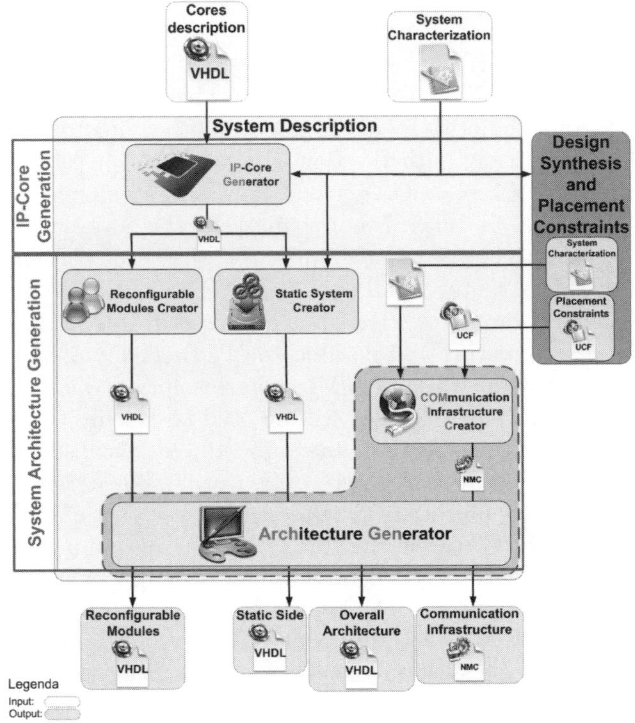

Figure 5.3: System description phase overview.

It is possible to identify four different sets of files used to describe the corresponding four basic architecture components: the reconfigurable side, the static one, the communication infrastructures, and the overall system architecture. In this context this system description phase has been organized into two different steps: the IP-Core generation and the system architecture generation. The former is the most application-oriented phase. At this point all the functionalities that have to be implemented in HW, both static and reconfigurable, are defined. The system architecture phase is used to define the overall system architecture. This architecture is composed of a set of application-independent components, e.g., the reconfiguration controller core[§] and the GPP, and a set of components that belong to the application that has to be implemented.

[§]The core that is responsible for the physical implementation of a reconfiguration, i.e., an ICAP controller.

5.3.1.1 The IP-Core generation

The aim of this phase is to build a complete IP-Core from its core logic. This task can be done manually or using automatic tools [135, 99, 186]. In the following we describe the *IPGen* tool [135], which is able to automatically create the desired IP-Core following three steps: *registers mapping*, *address spaces assignment*, and *signals interfacing*. The registers mapping step is needed because each core may have different (number, type, size) sets of signals, therefore this phase creates the correct binding between the register defined in the core logic and the one that has to be created inside the new IP-Core. The second step that has to be performed is the address space assignment. Each register, mapped to a standard signal, is now assigned to a specific address, allowing us to address a specific register through the address signals. The last step consists of the signals interfacing phase. Target bus signals are mapped to registers. After the execution of this sequence of steps, the IP-Core is ready to be bound to the target bus and has a proper interface.

Let us look in detail at how the design of an EDK-compatible IP-Core can be obtained. This *design* can be divided in two phases: the creation of the *core* and the implementation of the communication infrastructure. To allow rapid IP-Core generation, the second step has to be performed automatically by IPGen, which takes as input the VHDL file that implements the core and the name of the chosen communication architecture and produces as output the final IP-Core. The IPGen is composed of three main steps:

The input phase: this tool needs the VHDL description of the core and the indication of the chosen communication architecture.

The Reader: it reads, interprets, and saves all the information needed by the following step. The main operations performed by the reader are the following:

- Recognize the pattern of a VHDL entity declaration in the VHDL core description.
- Build the generics and I/O signal lists: when a generic is recognized, it is analyzed and its information is stored in the generics list. This action is repeated until the end of generics declaration. The signals list is built in the same way.
- Save the core entity name and the file path in two variables used by the following process.

The Writer: it writes the IP-Core VHDL description, performing the following actions:

- Accept as inputs the generics and I/O signals lists, the core's entity name and the core's path.

- Re-write the VHDL description of the core, because it is necessary in some cases to execute small changes to *clean* the core file in order to produce a correct implementation of the complete IP-Core.
- Create a stub VHDL file between the Core and the IP-Core VHDL description, which allows the input signals of the core to be written by the bus master, and the outputs to be read.
- Create the top architecture VHDL file, that is the final IP-Core, which contains the processing logic and the chosen interface.

Therefore, three output VHDL files are produced: the *cleaned* core, the stub, and the top IP-Core description. The resulting IP-Core structure is represented in Figure 5.4

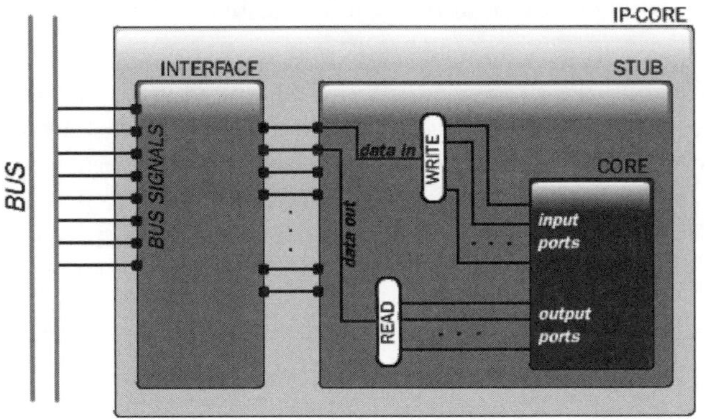

Figure 5.4: IP-Core final structure.

During the execution, if one of the two phases fails the tool is halted and an error message, useful to understand where and why the problem occurred, is generated. If the execution ends correctly the created IP-Core is ready to be plugged into an EDK system. It is important to specify that the tool is not a VHDL parser because it does not validate the analyzed code, except for the entity declaration syntax. An important feature of the tool is that the address decoding logic is automatically generated and included in the stub. Therefore, with this solution, while creating a hardware module, the designer does not have to deal with the core interconnections within the system, focusing his efforts instead only on the functionality itself. In other words, the core must not include any reference to the interfacing architecture.

The set of tests presented in Table 5.1 concern several types of components, starting from some small IP-Cores such as an IrDA interface to more complex examples, e.g., a complex ALU, a video editing core that changes the image coloring plane from RGB to YCbCr. Table 5.1 presents some relevant results, considering both the input core (the first row for each core) that represents the core logic, and the obtained component (the second row) that is the final IP-Core produced by the *IPGen* tool, [135]. For each one of them, the size in terms of 4-Input LUTs and the number of occupied slices are illustrated, both as absolute values and as the percentage with respect to the total size of the FPGA. Columns labeled with ar_i and d_i present the number of available resources (ar_i) given the minimum area constraints to implement the core as reconfigurable elements, and the density (d_i) computed as the ratio between the number of slices used to implement the core and the number of available resources, ar_i. The last column shows the time needed by *IPGen* to create the IP-Core.

Table 5.1: IPGen Tests

IP-Core	**4-Input LUTs**	**Ratio**	**Slices**	**Ratio**	ar_i	d_i	**Time (s)**
IrDA	15		11		136	0.081	
	146	9.73	103	9.36	136	0.758	0.045
FIR	273		153		272	0.562	
	308	1.13	173	1.13	272	0.636	0.058
RGB2YCbCr	848		913		952	0.959	
	1028	1.21	940	1.03	952	0.987	0.063
Complex ALU	1750		950		952	0.997	
	2089	1.19	1079	1.14	1088	0.991	0.071

On one hand the relative overhead due to the interface of the core logic with the bus infrastructure is acceptable, both for the 4-Input LUTs and for the occupied slices, especially when the core size is relevant. Therefore, from this point of view, the best case is to have a big design with few and small ports, while the worst one is to have a small core with many I/O signals. This result may suggest the use of bigger cores, for example, putting together in a unique module different functionalities. In system design, especially in a dynamic reconfigurable architecture, this solution has to be adopted carefully, for two main reasons: it implies a decrease in the degree of flexibility, which is one of the main features of modular-based design, and the time required by the reconfiguration becomes longer. Hence it is desirable to reach a reasonable trade-off between the size of the modules and the occupation of the overhead required by the communication logic.

In order to create the IP-Core, starting from input Core, IPGen needs an execution time that is almost constant and on average of 0.065 seconds. The generated IP-Cores will then be provided as input to the Reconfigurable Modules Creator or to the EDK System Creator tools, according to their nature, static or reconfigurable, as defined in the system characterization file.

5.3.1.2 System architecture generation

The main structure of a self-reconfigurable architecture consists basically of two parts: a static part and a reconfigurable one. This is valid for a single FPGA architecture [24, 57], as for a Multi-FPGA solution [109]. The reconfigurable side can be considered as a runtime customizable area due to a collection of reconfigurable modules (IP-Cores). Each possible combination of the static side with a different set of reconfigurable modules gives rise to a different configuration of the system. It is possible to consider, in a general view, that these configurations are used to create the bitstreams (a complete bitstream and a group of partial reconfiguration bitstreams) that will be used to set up the system (the complete bitstream) and to swap from one configuration to another one (the partial bitstreams).

The **reconfigurable modules creation** accepts as input the modules generated by *IPGen* and it provides as output the VHDL description of the corresponding reconfigurable modules. According to the experimental results we obtained, it has been possible to see that it does not affect the completion time of the system architecture phase. This is no longer valid for the **static system creation** stage. Input to this stage, in addition to the base EDK architecture used to define the core of the static side of the self-reconfigurable architecture, are the IP-Cores, provided by *IPGen*, that have been selected to be inserted in the static part of the architecture. The static system creator added to the base EDK architecture all the IP-Cores that have to be considered as static components aiming at defining the VHDL description of the static component of the desired architecture.

As shown in Figure 5.1, the next two stages of the *System Description* phase, even if included in this phase, come after the *Design Synthesis and Placement Constraints Assignment* phase, because they require placement information regarding the layout of the overall architecture.

One of the key aspects in designing a reconfigurable system is the definition of the communication infrastructure. As described in more detail in Section 7.2.1, Xilinx provides a specific hardware component, called bus macro [95, 98], to allow signals of reconfigurable functional unit to directly interact with the external hardware designs. Other solutions to create custom communication infrastructure for reconfigurable system can be found in the literature [76, 77, 115, 116]. In this section we present the ComIC, COMmunication Infrastructure Creator, a tool that has been developed to implement the **communication infrastructure creation** stage. It takes as input the placement constraints identified by the design synthesis and placement constraints assignment phase and the information regarding the communication protocol that has to be implemented, and it provides as output the corresponding macro-hardware that is used to implement the communication infrastructure. This tool uses the XDL language, like the works proposed in [76, 77], to model the desired communication infrastructure, like the one proposed in Figure 5.5. Once that the XDL file is generated, it is necessary to convert it in a

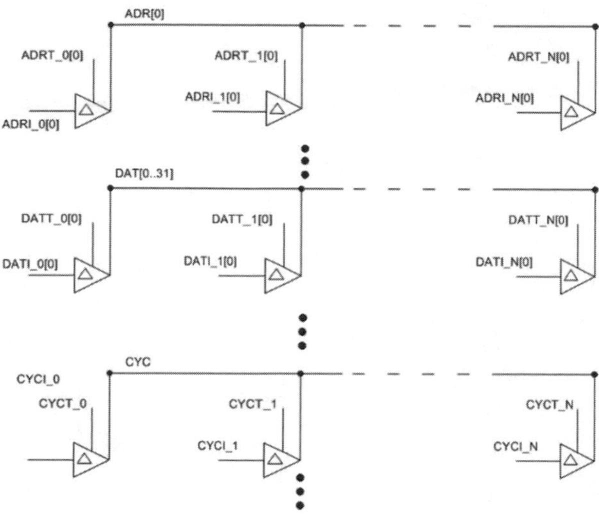

Figure 5.5: Custom bus macro layout.

macro hardware file (NMC). Once this operation is also done, the hard macro is ready to be used. An example of a custom bus macro, generated using ComIC, is shown in Figure 5.6. Finally, the **architecture generation** stage consists of the creation of the VHDL description of the top architecture where the static component, the communication infrastructure, and the necessary reconfigurable components are instantiated. This is achieved by analyzing the VHDL descriptions generated by the previous stages. Experimental results have shown that the most time-consuming stages are the static system creation and the architecture generation ones, therefore Table 5.2 reports the results of a set of experiments where the static side of the architecture has been designed using different kinds of processors, the PowerPC 405 and the Xilinx Microblaze.

The first column reports the main characteristics of the static part of the architecture under test. Let's take into consideration VP7MB1, meaning that the target FPGA chosen to implement the final solution is the Xilinx Virtex II Pro 7 (VP7) and that the processor instantiated on the static side was a Xilinx Microblaze. We use 1 or 2, just to characterize different static architecture, where different means that they have been defined using different sets of IP-Cores.

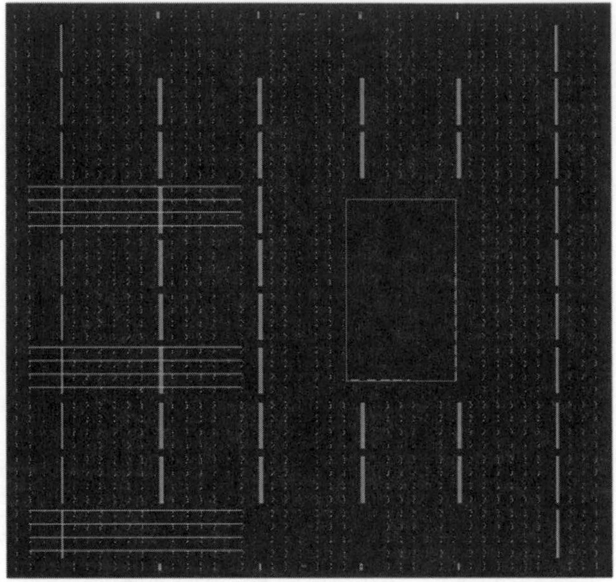

Figure 5.6: An FPGA view of a custom bus macro physical implementation.

Table 5.2: Architecture Generation Stage Time Requirements

	Parsing		Static		Top	
	(ms)	%	(ms)	%	(ms)	%
VP7MB1	13981	44	13666	43	3783	12
VP7MB2	16489	41	16490	41	6776	17
VP7PPC1	21339	47	18989	42	4964	11
VP7PPC2	22262	41	22901	43	8476	16
VP20PPC1	15413	42	16533	45	4841	13
VP20PPC2	17292	42	16898	41	7135	17
V4MB1	11631	42	11915	43	3937	14
V4MB2	13642	38	14755	42	6989	20

5.3.2 Design Synthesis and Placement Constraints Assignment Phase

The aim of this phase is the definition of the placement constraints that the reconfigurable architecture has to satisfy, such as the position of the reconfigurable slots or the physical location of bus macros. Given a set of reconfigurable modules m_i, $i = 1 \dots M$ and a set of time intervals t_k, $k = 1 \dots K$, the system configuration at each time is described through a binary relation $\varphi(i,k)$, i.e., a module $m_{\bar{i}}$ has to be configured and running at $\bar{k}$ if and only if $\varphi(\bar{i},\bar{k})$ holds. Given a time $\bar{k}$, $P(\bar{k})$ is defined as the set of all the modules being up and running at $\bar{k}$, i.e.,

$$P(\bar{k}) = \{m_i \, i = 1 \dots M | \varphi(m_i, \bar{k}) = 1\}$$

$P(\bar{k})$ will be referred to as the *static photo* of the reconfigurable system at instant $\bar{k}$.

A system, in order to be feasible, has to meet its resource constraints. Given

- a set of resources r_l with $l = 1 \dots L$
- $b_{m,l}$ the matrix containing the number of resources of type l needed by module m
- B_l the number of resources of type l available

The resource constraints at any time can be written as:

$$\sum_{i=1}^{M} \varphi(m_i, k) b_{m,l} \leq B_l \qquad \forall l, k$$

An area of the FPGA has to be assigned to each module belonging to static photo $P(\bar{k})$ and, in order to allow a modular reconfigurability, a set of area constraints has to be assigned. One of the aims of this section is to present an approach to solve the floorplanning problem for a single static photo optimizing a given objective function. One of the most remarkable works proposed in literature on the topic is [173]. In [173] the floorplanning problem of reconfigurable architectures is solved by optimizing the floorplanning of each static photo and then refining each floorplanning between two time-adjacent static photos using simulated annealing. The main drawback of such a technique is that the objective function is minimized on each single static photo and not globally on the entire set of static photos.

Currently, automated tools like PlanAhead [104] provided by Xilinx Inc. support modular floorplanning using brute force techniques that are often very time-consuming and thus require entire running days to optimize non-trivial designs. Even with such a brute force approach such tools do not support floorplanning for reconfigurable architectures, due to the fact that modules belonging to different static photos have to be considered at the same time.

The proposed approach exploits results of different placement assignment techniques, i.e., *game theory* based, in order to optimize a given objective function, but it can be easily extended to support global optimizations over several (and may be all) static photos. The Design Synthesis and Placement Constraints Assignment phase, as shown in Figure 5.7, needs as input the set of reconfigurable modules m_i written in a hardware description language such as VHDL or Verilog. Also $\varphi(m_i, k)$ and B_l are provided as input, the former depending on the architecture that is going to be implemented, the latter depending on the device on which the architecture is going to be deployed.

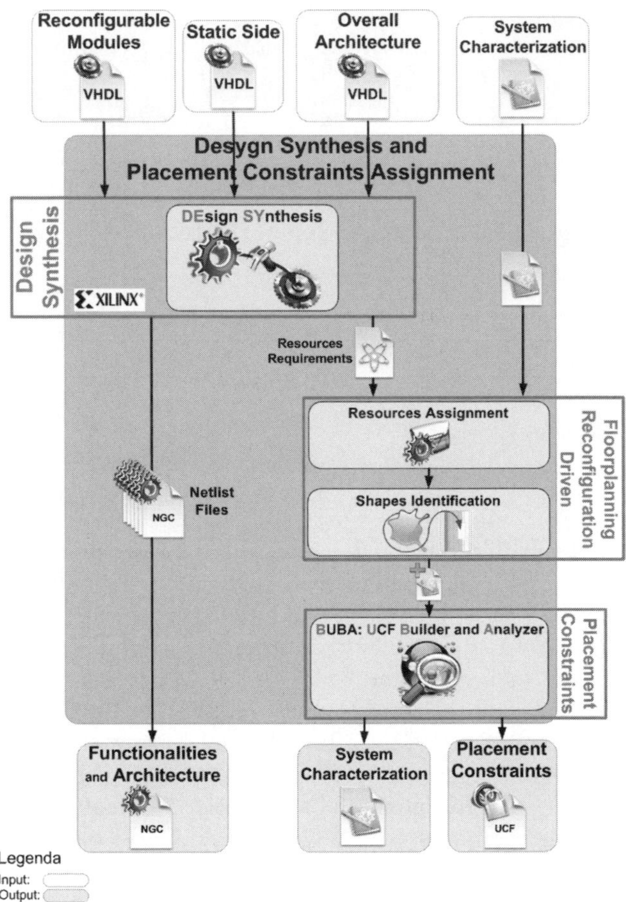

Figure 5.7: Design Synthesis and Placement Constraints Assignment Phase.

During the preliminary phases of the flow each module is synthesized in order to get the list of required resources. During this preliminary phase the previously introduced matrix $b_{m,l}$ is built up. According to φ and B_l an early feasibility check can be performed. The first phase of the proposed algorithm consists of the definition of a starting feasible floorplanning for each static photo $\varphi(\cdot, k)$. During the *Resource Assignment* phase all the available resources B_l are assigned to each $m_l \in P(\overline{k})$. Then each module goes through a *Shape Identification* phase: according to the objective function a shape for each module m_l is proposed in order to optimize the given objective function. In this phase we look for an optimal floorplan shape for each module, but adopting optimal shapers for each module of a static photo $\varphi(\cdot, \overline{k})$ is not a feasible solution, hence the optimal shape is used just as a starting point for the core of the algorithm.

All of the state-of-the-art algorithms face the floorplanning problem by optimizing at most one single static photo at a time. The improvement given by the proposed work flow is that the entire architecture is optimized with respect to the objective function considered. The core of the optimization process is given by the last phase generating the final placement constraints.

The aim of this latter stage is the identification of the placement constraints that will be used to implement each module. The previous phase generated an optimal shape, but the problem that still needs to be solved is the identification of the placement constraints. At the end of the flow a set of area constraints is provided, one set for each static photo composing the design. In the next section further details of each phase are proposed.

5.3.2.1 Design Synthesis

This stage is used to synthesize each functionality description provided to this stage using VHDL or Verilog, to estimate the resources that will be required to define the corresponding configuration code.

5.3.2.2 Reconfiguration Driven Floorplanning

This stage aims at the definition of the area constraints for each configuration code. Once the estimation for the resources required by each configuration code are provided by the design synthesis phase, it is possible to identify a floorplanning constraint that takes into consideration both the resource requirements and the constraints introduced by the reconfigurable scenario (i.e., working with a Xilinx device, a width constraint multiple of 4 slices [95, 98]).

In order to define this area constraint it is possible to use the fragmentation index, which represents a percentage of wasted space (relative to the size of a region). It can be computed using the following formula:

$$\frac{(\#region\_slices) - (\#occupied\_slices)}{\#region\_slices} * 100 \tag{5.1}$$

In the fragmentation index, see Figure 5.8, the *#region_slices* parameter represents the number of slices available in the area defined by the area constraint into the UCF, User Constraint File, while the *#occupied_slices* value defines the actual number of slices used by the physical implementation of the core into its assigned area. To obtain the *#occupied_slices* for each module

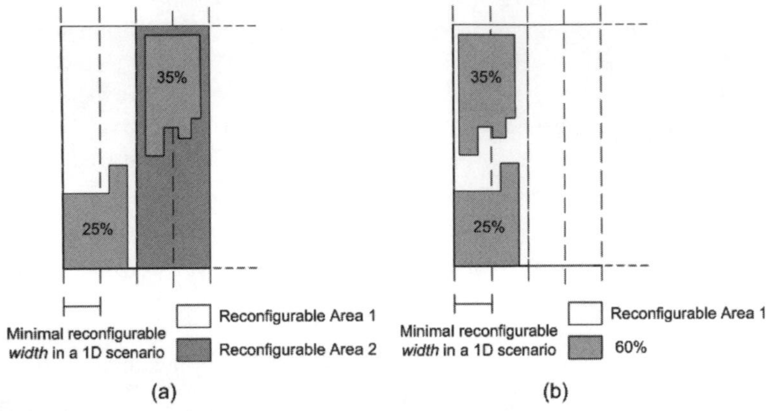

Figure 5.8: Working in a 1D reconfigurable scenario it is possible to see how the fragmentation index can be used to drive the placement of the reconfigurable modules that have to be implemented. (a) In this context we have to different reconfigurable regions, both with a high fragmentation index. (b) It presents a scenario where both the components are placed in the same are to minimize the fragmentation index.

the corresponding VHDL description has to be synthesized¶. The floorplanning reconfiguration-driven stage provides as output an area constraint aware of all the constraints introduced by the reconfiguration scenario.

5.3.2.3 Placement Constraints

The aim of this stage is the identification of the placement constraints that will be used to implement each configuration code. The floorplanning reconfiguration driven stage provides a set of feasible area constraints, but the problem that still needs to be solved is the identification of the placement constraints taking into consideration the fact that those configuration codes are not configured as single core on the reconfigurable device but have to share the reconfigurable resources with other configuration codes. The UCF Builder and Analyzer (BUBA) tool accepts as input the starting area solutions computed by the previous stage, a static scheduling of the application, and the information regarding the reconfigurable device that has to be used to implement the desired design. All this information is provided in the *system characterization* file. Due to these parameters BUBA tries to assign, in a greedy way,‖ the placement constraints to each module, trying to minimize the

¶The accuracy of the *#occupied_slices* value affects dramatically the quality of the placement.

‖We are now working on an new version of this framework, which is based on a non-greedy approach.

number of reconfigurations [72]. This is possible due to the fact that a module can be requested to be executed at different times and not only once. In such a scenario there might be a placement solution able to keep configured this configuration code on the reconfigurable device, without affecting the quality of the schedule, without having to reconfigure the same module twice or more just because it is no longer available on the reconfigurable device. Once the new placement constraints are defined this information is stored in the *system characterization* file and in the UCF file and provided, respectively, to the system architecture generation stage in the *System Description* phase to implement the correct communication infrastructure and to the context creation stage in the *System Generation* phase.

A *game theory*-based technique is now under development. The idea is to consider each core as a *player* in the game, where the objective of each player is the best placement assignment [141]. The set of all the players identifies the *society*. In this context the best solution for the society may not be equal to the best assignment of each core. This means that a placement assignment solution for a system corresponds to an equilibrium.

5.3.2.4 Validation results

This section describes the solution space of the proposed approach. In order to verify the proposed flow, several components and architectures have been implemented, and the proposed approach will be exemplified through a SDRAM memory controller.

Several metrics can be used in order to rate the floorplanning of a module, our flow considers three of them:

- the fragmentation factor
- the length of critical path
- the average length of the 10 longest paths

Previously it was underscored that one of the most critical steps to obtain a good result consists of the choice of the initial floorplanning for each module.

A single objective function has been defined as

$$\Theta = A\Phi^2 + B\Big(\frac{\tau_{\max}}{\max_{\text{design}}\{\tau\}}\Big)^2 + C\Big(\frac{\tau_{\text{10longest}}}{\max_{\text{design}}\{\tau\}}\Big)^2 \tag{5.2}$$

In order to make Θ the sum of parts belonging to the interval $[0, 1]$ the time components have been scaled using the longest critical path. Hence, if the considered module is responsible for the longest critical path then Θ will have a normalized time component equal to 1, and any variation to the critical path due to the floorplanning will be appreciated more than any other improvement obtained, i.e., decreasing fragmentation.

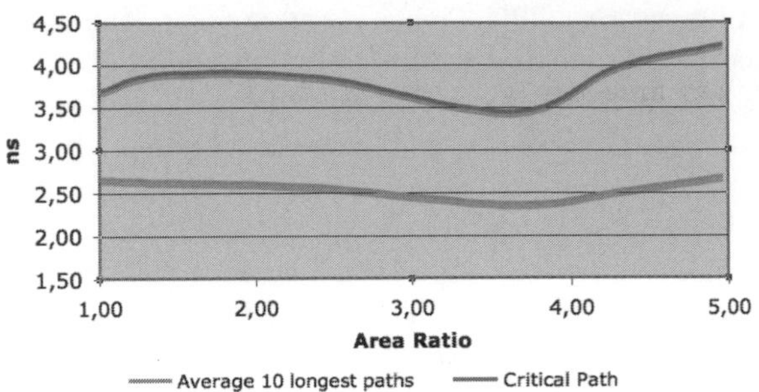

Figure 5.9: Critical path variation with respect to a variation of assigned square area.

The square power is used to assign a higher weight to higher components, i.e., it is better to gain a decrease in a long data path than a decrease in a low fragmentation rate. Furthermore, parameters A, B, and C can be specified in order to better weigh the importance of each component.

Figure 5.9 plots the delay along the critical path and the average of the 10 longest paths with respect to the square assigned area. On the x axis there is the area ratio with respect to the smallest feasible solution, i.e., an area ratio equal to 2 means having assigned a square area twice the minimum required. Furthermore increasing the area results in a routing becoming less knotty but, without introducing further constraints to the placement and routing tool, the logic becomes spread and the interconnections become longer and total latency increases.

Such conclusions also can be drawn for other components of objective function Θ, hence each component of Θ has a parabolic-like shape and a local search algorithm can be applied to find an area constraint that minimizes Θ.

Due to its algebraic definition, Θ is given by the sum of functions that can be approximated by parabolic functions along wide intervals, hence Θ also can be approximated locally by polynomial functions. Results of Lagrange multipliers theory can be applied to find the singularity points of the function. That is why the proposed approach adopts an optimized local search method driven by the discrete gradient of Θ.

Considering the scenario proposed in Figure 5.10 we assigned different area constraints to the SDRAM memory controller. Table 5.3 presents the results of the corresponding technology mapping. In order to describe the shape of the assigned area, a coefficient given by the ratio between height and width of the area has been associated with each feasible solution. Table 5.3, presenting the SDRAM memory controller synthesis data corresponding to different area constraints assignment, shows that an increase in the assigned area does not always result in a better timing result. This is due to mapping tools (such as the one used in the Xilinx ISE environment) that tend to spread logic along the assigned area without caring, unless explicitly specified, about global timing.

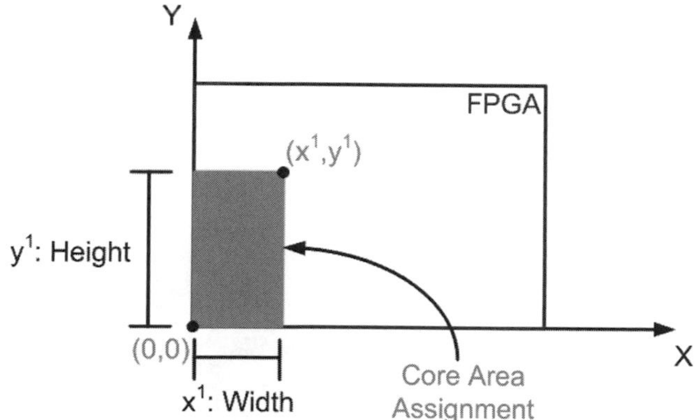

Figure 5.10: Overview of the system used to assign the height and the width of the area constraint of a generic module.

Table 5.3: SDRAM Memory Controller Synthesis Results.

Width	Height	Height/Width	Longest Delay	Average Conn. Delay Worst 10 Nets	Fragmentation Index
13	13	1.00	3.90	2.75	0.85
17	17	1.00	3.87	2.58	0.49
21	21	1.00	3.45	2.38	0.32
23	23	1.00	3.99	2.52	0.27
9	21	0.43	2.94	2.39	0.76
9	17	0.53	3.20	2.39	0.94
9	25	0.36	4.02	2.67	0.64
17	9	1.89	3.69	2.79	0.94
21	9	2.33	3.88	2.54	0.76
25	9	2.78	4.04	2.42	0.64
5	31	0.16	3.74	2.93	0.93

The proposed work-flow generates an optimal floorplanning for each module exploiting the described mathematical approach. Each set of modules belonging to the same static photo goes through a final floorplanning phase using a greedy algorithm to schedule as best as possible the entire static photo.

5.3.3 System Generation Phase

The previous two phases produce all the necessary files (i.e., HDL descriptions, UCF file, macro-hardware definitions, etc.) to describe the desired system, both in its static and reconfigurable side. The last phase of the flow is the *System Generation* phase; see Figure 5.11.

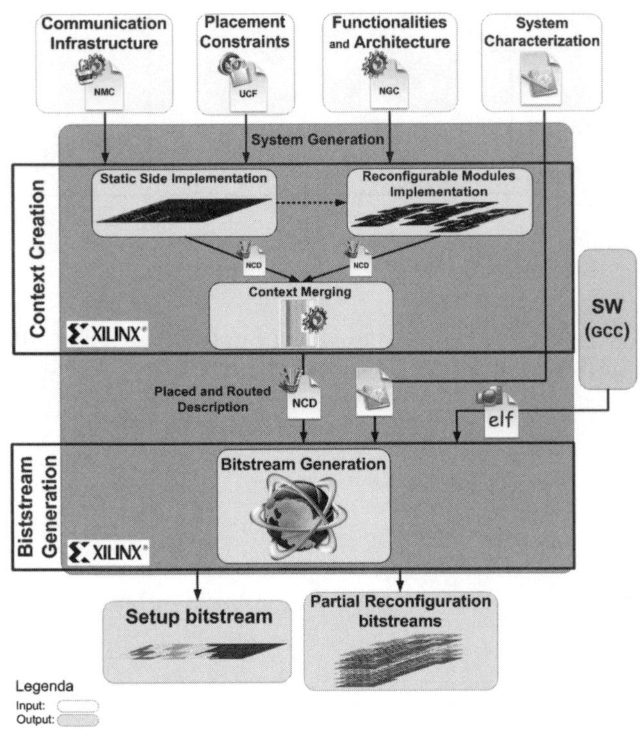

Figure 5.11: System Generation phase overview.

The System Generation phase can be used to support both the Xilinx Early Access Partial Reconfiguration-based [98] and the Module-based [95] reconfigurable architecture design flow. This phase has been divided into two different stages: the **Context Creation** and the **Bitstream Generation**.

The **Context Creation phase** has been organized into three different stages: Static Side Implementation, Reconfigurable Modules Implementation, and Context Merging. The first one accepts as input the HDL files** generated during the system description phase and the placement information defined

**NMC and NGC files.

via the design synthesis and placement constraints assignment phase. The aim of this stage is the physical implementation of the static side of the final architecture. The placement information of this component will be provided as input to the Context Merging phase of the final architecture. A second output, working with the Xilinx EAPR flow, is represented by the information of the static side components that are placed into the reconfigurable region, i.e., routing, CLBs usage. This information is stored in the *arcs exclude* file, and it is represented by the dotted line, in Figure 5.11, used to bind the Static Side Implementation and the Reconfigurable Modules Implementation stages. The *Reconfigurable Modules Implementation* stage needs as input the placement information for each module and the corresponding VHDL files defined during the previous two phases and the *arcs exclude* file. This stage defines the physical implementation of each reconfigurable component that has to be passed as input to the hardware merging phase. It is composed of three different steps: the NGDBuild, the mapping, and the final place and route stage. The Reconfigurable Modules Implementation stage needs to be executed for each reconfigurable component. Finally, the *Context Merging* stage produces as a result the merging of the outputs produced by the two previous stages. Table 5.4 reports the average execution time for each phase composing the Context Creation stage. We report the results for all the devices used in our tests: Xilinx Spartan3, Virtex II Pro, and Virtex 4.

Table 5.4: **Context Creation** Stage Execution Times

Phase	S200 (s)	VP7 (s)	FX12 (s)
Static Side Implementation	133.86	224.91	1060.28
Rec. Modules Implementation	110.72	165.69	249.72
Context Merging	466.19	852.13	1407.36
Total	710.77	1242.73	2717.36

The **Bitstream Generation** stage consists of the generation of a complete bitstream, used to start up the system on the FPGA, and two partial bitstreams for each reconfigurable functional unit that has to be implemented: one is used to configure it over an empty module and the other one to restore the empty configuration.

Summary

This chapter, building on the information provided in Chapters 3 and 4, describes different solutions proposed in the literature to design and implement dynamic reconfigurable systems. The main objective of this chapter is to describe a general and complete design methodology that can be followed as a guideline for designing reconfigurable embedded systems. The idea behind the methodology presented in this chapter is based on the assumption that it is desirable to implement a flow that can output a set of configuration bitstreams used to configure and, if necessary, partially reconfigure a standard FPGA to realize the desired system. The proposed workflow aims at designing a complete framework able to support different devices (i.e., [102, 101]), multi-FPGA architecture (i.e., the RAPTOR2000 system [109]), different reconfiguration techniques ([95, 98]), and type of reconfiguration (i.e., internal or external, mono and bi-dimensional) that allow a simple implementation of an FPGA system specification, exploiting the capabilities of partial dynamic reconfiguration. Regarding the software side of the desired solution, it can be developed either as a standalone application or with the support of an operating system, such as Linux, a solution that was presented in Chapter 4.

Exercises and Problems

1. Which is the difference between modular-based and difference-based designs?

2. List as many novelty contributions introduced by the Xilinx EAPR design flow as you can.

3. List at least three different generic design flows for partially reconfigurable systems.

4. What is the main objective of the IP-Core generation phase?

5. What is a bus macro and what is it used for? Is there only one kind of bus macro? If not, explain why.

6. Describe the placement constraints assignment phase in your own words.

7. What is the fragmentation index?

8. **Project Theme**: Describe the architectural details of a typical reconfigurable system that can be obtained with the proposed design flow.

List a couple of critical aspects that you have to take into account during the design phase.

6

Reconfigurable System Verification

Besides the characteristic behavior variations between hardware and software, a reconfigurable system design has an additional dimension of complexity. Hardware and/or software may be reconfigurable at runtime, which leads to not only behavior changes of fixed components, but also architectural changes that might drastically alter the system behavior. Further, conventional CAD or simulation tools do not support runtime reconfiguration. It is a basic assumption in currently available tools that the system architecture is fixed statically at design time. However, reconfigurable systems violate this basic assumption and thus it is not feasible to use conventional tools and methodologies to verify them. Verification methods and tools specifically targeted at reconfigurable systems exist but are still in the development stages. In this chapter, we first highlight some of the tools and methods that can be used to verify reconfigurable systems. We will then focus on a design space exploration framework called *Perfecto* [90].

6.1 System-Level Verification Techniques

System-level verification is important for hardware-software systems such as SoC and embedded systems. However, it becomes even more important for reconfigurable systems because it is necessary to check if the system behaves correctly after reconfiguration. Formal verification is one method that can ensure correct system behavior.

6.1.1 Formal Verification

The SymbC tool [26] used in the Symbad design flow for reconfigurable Systems-on-Chip (SoC) is a typical example of how hardware and software are co-verified. SymbC takes as input C programs and FPGA configurations and verifies them formally. However, it is restricted to checking a fundamental consistency property: "Each time the software requires a hardware resource of the reconfigurable part, this resource is actually available." SymbC generates a list of potential problems if the consistency verification fails. The C programs are pre-processed by gcc and abstracted before verification. The tool was applied to a face recognition system with image preprocessing such as Bayer filter and face location based on Hough transform. Eight incorrect reconfigurations of the hardware were detected by SymbC.

The Xilinx COREGen tool can be used to design reconfigurable cores for the Xilinx FPGA platforms. Singh and Lillieroth [172] proposed the dynamic generation of the question formula for a new dynamically reconfigured core generated by COREGen and then using the theorem prover called *Prover* to ensure the dynamically calculated circuit is correct. When a system wishes to perform architecture reconfiguration, it first updates its internal model of the circuit and submits it to Prover for re-verification against the new specification. Using lazy evaluation it can avoid re-proving parts of the circuit that change, and the proof effort is concentrated on the reconfigured parts.

Fey et al. [67] proposed another approach to formal verification of reconfigurable systems. The configurations of a functional block in a reconfigurable system were abstracted fully into a set of relations. Then, *Binary Decision Diagrams* (BDDs) were used to represent the relations. Finally, circuits were generated from the BDDs.

6.1.2 Language Approach

Different kinds of architecture description languages (ADL) and temporal logic can be used to specify and analyze properties of reconfigurable systems. ADLs [6, 203] provide constructs to model the system architecture. Allen and Garlan [6] formalized architectural connections that can have different roles and a glue stating the interactions between the roles. Automated compatibility checks allowed the connectors to be well defined. Thus, this language can be used to reason about reconfigurable systems that might have dynamic changes in architectural connections. A simple example of this language is as follows, where $\Box$ and $\boxdot$ represent deterministic and non-deterministic choice between the left and right operands, respectively, $\surd$ implies `success`, the symbols ? and ! represent data input and output, respectively, and the bold words are keywords.

connector Service =

role Client = request!x $\rightarrow$ result?y $\rightarrow$ Client $\Box\surd$

role Server = invoke?x $\rightarrow$ return!y $\rightarrow$ Server $\boxdot\surd$

glue = Client.request?x $\rightarrow$ Service.invoke!x

$\rightarrow$ Service.return?y $\rightarrow$ Client.result!y $\rightarrow$ **glue** $\boxdot\surd$

The server role repeatedly accepts an invocation and then returns, or it terminates with success instead of being invoked. The client role either invokes the service or terminates. The glue process coordinates the behavior of the two roles by indicating how the events of the roles work together. The glue allows the client role to decide whether to invoke or to terminate and then sequences the remaining three events and their data. This example illustrates how the connector description language is capable of expressing the conventional notion of providing and using a set of services.

Further, Wermelinger et al. [203] proposed an imperative language that can be used to specify categorical diagrams and dynamic reconfiguration through algebraic graph rewriting. This language supports component creation, component refinement, connector creation, node removal, node query, and variable assignment. The language uses category theory as a semantic foundation for both configurations and reconfigurations. It has design characteristics for computation, declarative characteristics for constraints, and operational characteristics for reconfiguration. Following are some examples of how components can be created, refined, and connected. The words in bold are keywords. The keyword **create** is used to create a new component called $Node_1$ of type *Design* with conditions such as $l_i := Exp_i$. The component $Node_2$ is created as a refinement of the component $Node_1$ with conditions $l_i := Exp_i$. A connector called *Node* is created by applying refinements to $Node_i$.

$[Node_1 :=]$
create *Design*
with $l_1 := Exp_1 \parallel l_2 := Exp_2 \ldots$

$[Node_2 :=]$
create $Design_2$ as $[Refinement(]Node_1[)]$
with $l_1 := Exp_1 \parallel \ldots$

$[Node :=]$
apply $Connector([Refinement_1 \rightarrow]Node_1, \ldots)$
with $l_1 := Exp_1 \parallel l_2 := Exp_2 \ldots$

Temporal logic-based language such as that proposed by Aguirre and Maibaum [2] can be used not only to specify architecture but also reason about its correctness. This language can be used to specify classes, associations between classes, and subsystems. Each subsystem specifies which instances of each of the classes are included in the subsystem. Properties can be proved for both classes and subsystems. For example, we can prove that for the lifetime of a multicore system, some particular resource is only occupied by a processing core C.

$$\forall x : Core(x) \rightarrow (x = C)$$

6.2 Hardware-Software Coverification

The possible interactions among hardware and software components in a system have always been a major bottleneck in verifying hardware-software systems. Now, with the added dimensionality of static or dynamic reconfiguration, the bottleneck has become worse. For example, with fixed hardware and software, there is no issue about the existence of either when a communication is desired. At most, the communication protocol is violated by either components or the performance is poor. However, with dynamic reconfiguration, even the existence of components becomes a question for further investigation. Whenever a component A needs to communicate with another component B, it is necessary that B exist in the system. If there is no support from the OS4RS that B will be brought into the system on the fly whenever required, then the problem exists and must be solved. Even if the OS4RS provides support for dynamic module instantiation as described in Chapter 4, it is still required to verify that a required module exists and is usable in the system. Hardware-software coverification will try to detect all such errors in the interactions between hardware and software, even under reconfiguration. In the rest of this section, we will discuss hardware-software cosimulation and prototyping.

6.2.1 Hardware-Software Co-Simulation

Hardware-software cosimulation is a technique by which we can verify not only the individual components but also the communication among them. A typical cosimulation framework consists of a CPU simulator and a hardware simulator working together to evaluate a hardware-software system. The bottleneck in cosimulation has been the large difference in simulation speed between hardware and software. Hardware simulation is usually several orders of magnitude slower than software simulation if software is simulated at the instruction level and hardware at the register-transfer level.

The situation becomes all the more intricate when the hardware is reconfigurable because we now need not only to simulate the behavior of components, but also to simulate the reconfiguration of the system. Reconfiguration of a system can be simulated in different ways depending on the simulators used. If a conventional hardware simulator such as ModelSim is used then one could simulate a *Reconfigurable Unit* (RU), in which reconfigurable hardware could be configured. Enzler et al. [61] designed an RU model consisting of computational units, routing network, and circuitry for reconfiguration and I/O interfaces. This RU model was then simulated at cycle-accurate level using the ModelSim hardware simulator. The SimpleScalar CPU simulator was used to simulate the software that was written in the C programming language. RU configurations (bitstreams) were retrieved by the CPU and then sent to the RU. A library was provided to download and switch between RU contexts and to transfer data between CPU and RU. The CPU and the RU were synchronized using a down-counter. This cosimulation environment was applied to FIR filter partitioning and mapping, which consisted of eight FIR subfilter contexts.

6.2.2 Hardware-Software Prototyping

Since hardware-software cosimulation is basically performed in a fully software environment, it could be relatively slow. An alternative is to use a hardware-software prototyping platform, which consists of hardware accelerators and actual CPU and FPGA chips on a development board. This type of prototyping is often used in the design of embedded systems. For reconfigurable systems, the availability of development platforms would also play the same role and accelerate the design and verification phases.

A typical example of a prototyping platform is the Rapid-Prototyping System Raptor 2000 platform developed by the Heinz Nixdorf Institute, University of Paderborn, Germany. It consists of a motherboard and up to six application-specific modules (ASM). The motherboard provides a communication infrastructure for the ASMs, a PCI bus link to a host computer, and two Complex Programmable Logic Devices (CPLDs) to perform bus arbitration, memory management, and error detection. There are four kinds of services in Raptor 2000, namely Communication, Configuration, Management, and Monitoring.

In communication, there are two buses including a local bus and a broadcast bus. The local bus interconnects the ASMs for internal as well as external communication. The broadcast bus is used to simultaneous communication with all modules by any module.

In configuration, each ASM has a Xilinx Virtex FPGA device such that it can be configured with the required function before use. Raptor 2000 implements the fastest parallel and the serial JTAG configuration modes for configuring all the FPGAs. The configuration manager is implemented in hardware and supports partial reconfiguration and self-reconfiguration by any device in the system.

In management, a manager is used to control the access to intra-system resources such as the local bus, bus initialization for each master, and the assignment of local bus addresses to the different devices and ASMs.

In monitoring, system crashes such as illegal memory accesses are avoided through a monitoring unit.

A complete software environment is provided to users by Raptor 2000. A graphical user interface is also provided. All important registers of the PCI bus bridge, the MMU, and the configuration manager can be visualized and modified.

6.3 Reconfigurable System Simulation Frameworks

Since the design of reconfigurable systems mainly differs from conventional systems in the dynamic characteristics of changing system architecture and behavior, conventional methods such as cosimulation and prototyping can only verify one or a few pre-determined system configurations. It becomes especially cumbersome to use conventional methods and tools when there is a large number of possible configurations for a system. For example, using the hardware-software cosimulation environment or the Raptor 2000 platform, we can only simulate or emulate one configuration at a time. Another system configuration can be evaluated only after reconfiguring the system and restarting simulation or emulation.

As a solution to the above issue, system-level performance evaluation and design space exploration frameworks have been proposed. These frameworks can automatically evaluate all possible configurations at the expense of a small loss in inaccuracy. For example, most of such frameworks perform verification at the *transaction level*, which is fast but might not be as accurate as register transfer level or instruction level.

In creating such a framework, there are several issues to be resolved as follows. We will present solutions to the issues based on Perfecto [90], a design space exploration framework for reconfigurable systems.

- *Architecture Model*: How must one define a parameterized architecture model such that users can configure it by simply setting the parameters?
  - Section 6.4.1 gives a formal definition of the reconfigurable architecture model used in Perfecto.
- *Application Model*: How must one define an application model such that a generic application can be executed on the architecture model?
  - Section 6.4.2 gives a formal definition of the application model used in Perfecto.
- *Design Algorithms*: What kind of design algorithms must be supported by the framework such that the execution of the application model on the architecture model can be thoroughly analyzed?
  - Partitioning, scheduling, and placement algorithms are considered in Perfecto as described in Sections 6.4.3, 6.4.4, and 6.4.5, respectively.
- *System Evaluations*: What kinds of system features must be evaluated such that they are of use to system designers?
  - Section 6.4.6 gives the 4 task features and 5 partition features that are evaluated in Perfecto.
- *Guidelines*: In what ways can a framework guide designers in choosing the right design alternatives that meet user requirements?

– Perfecto helps designers in three ways: (a) making intelligent partition decisions, (b) optimizing performance, and (c) evaluating task placements. Examples are given in Section 6.4.7 to illustrate these guidelines.

Perfecto is based on SystemC [142], an IEEE 1666 standard system-level modeling language supporting both software and hardware specifications. Executable specification with simulation is an added benefit of SystemC, which is rapidly becoming the language of choice for system-level design. This is partly due to the fact that all large EDA vendors support or plan to support SystemC in their tools. Perfecto takes full advantage of unique features of SystemC such as built-in simulation, transaction-level modeling, software-hardware modeling, communication modeling, and performance evaluation.

6.3.1 Overview of Frameworks

Being a simulation-driven system description language, the IEEE 1666-2005 SystemC standard language has been used for design space exploration at the system level in several application domains such as embedded software, SoC, multicore systems, and reconfigurable architectures. The *Transaction-Level Modeling* (TLM) interfaces supported by SystemC not only make simulation possible at different abstraction levels, but also accelerate simulation, thus enabling rapid design space exploration. Further, due to the C++-based design, SystemC can be used to model both hardware and software. All these features make SystemC a very suitable language for exploring and evaluating reconfigurable system architectures and applications. Nevertheless, SystemC is still restricted in its capability to model dynamic reconfiguration because run-time instantiation of `sc_module` and dynamic binding of `sc_method` and `sc_thread` are not allowed in the current IEEE 1666 version of SystemC. Nevertheless, it was shown by Rissa et al. [158] that SystemC can achieve a much faster simulation speedup compared to traditional HDLs.

SystemC has been used for modeling and exploring reconfigurable system architectures in some previous work such as the ADRIATIC project [147, 151, 188] and the SyCERS framework [162]. The ADRIATIC project used SystemC to model dynamically reconfigurable systems by introducing a *Dynamically Reconfigurable Fabric* (DRCF) [147]. Reconfigurable components are all mapped to a DRCF that is generated from a template, which contains a context scheduler, an instrumentation process, and a multiplexer that routes data transfers to correct instances. Interfaces and ports of reconfigurable components are all added to the DRCF. Bus cycle accurate performance evaluations such as reconfiguration delay, context size, and computation delays can be obtained. The limitations are as follows. All components mapped to DRCF must be at the same level of hierarchy. Partial configuration was not supported because DRCF uses context switching to change between the different reconfigurable functions. DRCF was later extended into *Dynamically*

Reconfigurable Co-processors (DRC) [151], which support partial reconfiguration using a configuration scheduler and an input splitter. However, only reconfiguration latencies were evaluated. Energy consumptions were not modeled in DRCF and DRC and the support for the design space exploration of system architecture alternatives was limited.

SyCERS [162] is also a SystemC-based framework for the design space exploration of dynamically reconfigurable systems. Partial reconfiguration is supported. Instead of static binding of multiple functions through a multiplexer as in DRCF or through an input splitter as in DRC, SyCERS uses function pointers that are changed at run-time to model dynamic reconfiguration. SyCERS allows users to model the application through a fixed set of interfaces and to model the system architecture using several TLM black boxes constituting the YaRA architecture. SyCERS uses `sc_thread` to model a reconfigurable component and uses `sc_mutex` to synchronize configurations and executions. Elaborate details on the configuration process and the configuration controller are the focus of SyCERS, which can help a designer to decide on a more optimal architecture for a particular application, in terms of the number of black boxes. SyCERS does not focus on how hardware-software partitioning, scheduling, and placement are performed for an architecture/application combination.

The Perfecto framework [90] is more similar to SyCERS because Perfecto also tries to find a match between an application model and a reconfigurable architecture TLM model. The difference from DRCF, DRC, and SyCERS is that instead of fixed algorithms, Perfecto evaluates partitioning, scheduling, and placement algorithms along with an architecture and an application. Further, in Perfecto, design space exploration is automatically performed by providing an interface to random task graph generation, by evaluating multiple partitionings of the system, by detecting performance bottlenecks, and by evaluating the placement of all reconfigurable tasks in each partitioning. A designer can choose the best architecture by referring to the partition evaluations in Perfecto. Section 6.4.6 presents more details on how Perfecto performs design space exploration.

6.4 Perfecto Framework

Design space exploration and performance evaluation are extremely important but difficult for reconfigurable systems due to the complex dynamic nature of such systems and the multitude of combination possibilities in hardware-software partitioning, reconfigurable hardware scheduling, and reconfigurable hardware placement. Moreover, scheduling and placement must be concurrently considered for a feasible design solution. Perfecto is a framework proposed for integrating the design algorithms and for the design space exploration of dynamically reconfigurable systems through performance evaluation.

As shown in Figure 6.1, the design flow for dynamically reconfigurable systems is divided into two phases, namely a front-end design and a back-end design. The front-end design phase takes an architecture model and an application model, which might be derived from user-given system specification, and then generates three kinds of tasks, namely hardware task, reconfigurable hardware task, and software task. The back-end design phase synthesizes the three kinds of tasks, performs more detailed hardware-software co-simulation and then implements the full system using back-end tools such as hardware synthesis tool, software compiler, gate-level simulator, and power estimation tool. Perfecto nicely fits into the front-end design phase as the main tool for design space exploration and performance evaluation.

As shown in Figure 6.2, Perfecto takes two inputs, namely an architecture model and an application model, which are defined later in Sections 6.4.1.4 and 6.4.2, respectively. Hardware-software partitions are then generated by Perfecto. For each partition, the scheduler in Perfecto schedules the reconfigurable hardware and the software tasks. It is assumed here that the fixed hardware accelerators are part of the dynamically reconfigurable system architecture and thus not scheduled by Perfecto. Then, Perfecto simulates the execution of the software tasks on a processor, places the hardware tasks in the reconfigurable logic, and simulates their execution on the DRCL. Finally, after all task executions in all partitions are simulated, Perfecto generates several reports for the system designer, including partition evaluations, bus access conflicts, and real-time placement information, which are described in Section 6.4.6.

In the rest of this section, we will first describe the two inputs of Perfecto, namely the reconfigurable architecture model and parameters in Section 6.4.1 and the application model and task parameters in Section 6.4.2. Later, in Sections 6.4.3, 6.4.4, and 6.4.5, respectively, we will also describe the partitioning, scheduling, and placement algorithms that were implemented into Perfecto for illustration purposes. Designers can always change these algorithms or use the real-time information provided by Perfecto to construct a suitable algorithm. It must be noted here that the algorithms themselves are not the focus of design; rather, it is the evaluation framework design and how it can benefit designers. The algorithms will be the focus of design when we are designing an operating system for reconfigurable systems.

6.4.1 Reconfigurable Architecture Model

For performance evaluation, we need a basic architecture model of the target dynamically reconfigurable system. As shown in Figure 6.3, the basic reconfigurable system architecture model in Perfecto consists of a processor model, a memory model, a function ROM model, a bus model, an arbiter model, and a dynamically reconfigurable logic (DRCL) model. The SystemC design lan-

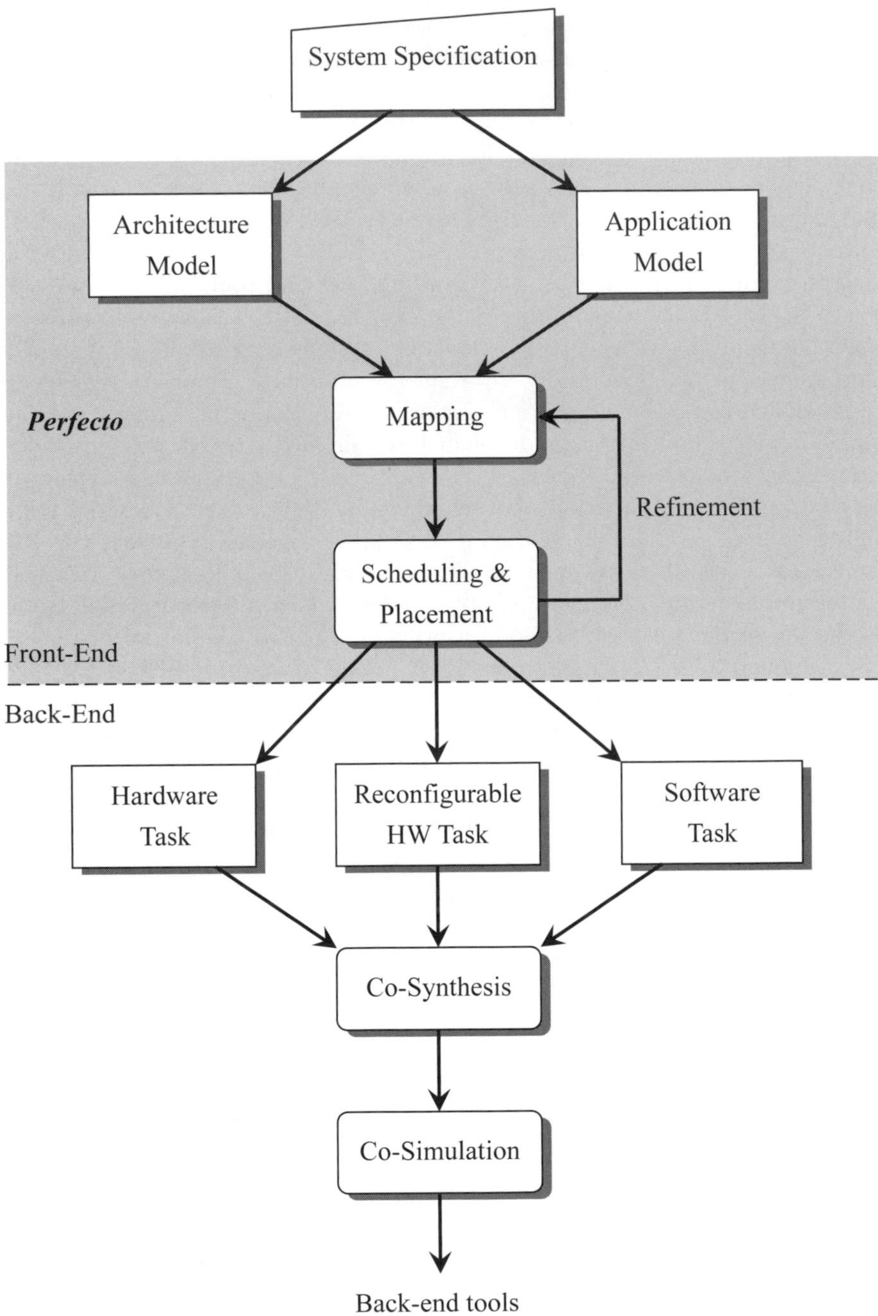

Figure 6.1: Design flow of dynamically reconfigurable systems.

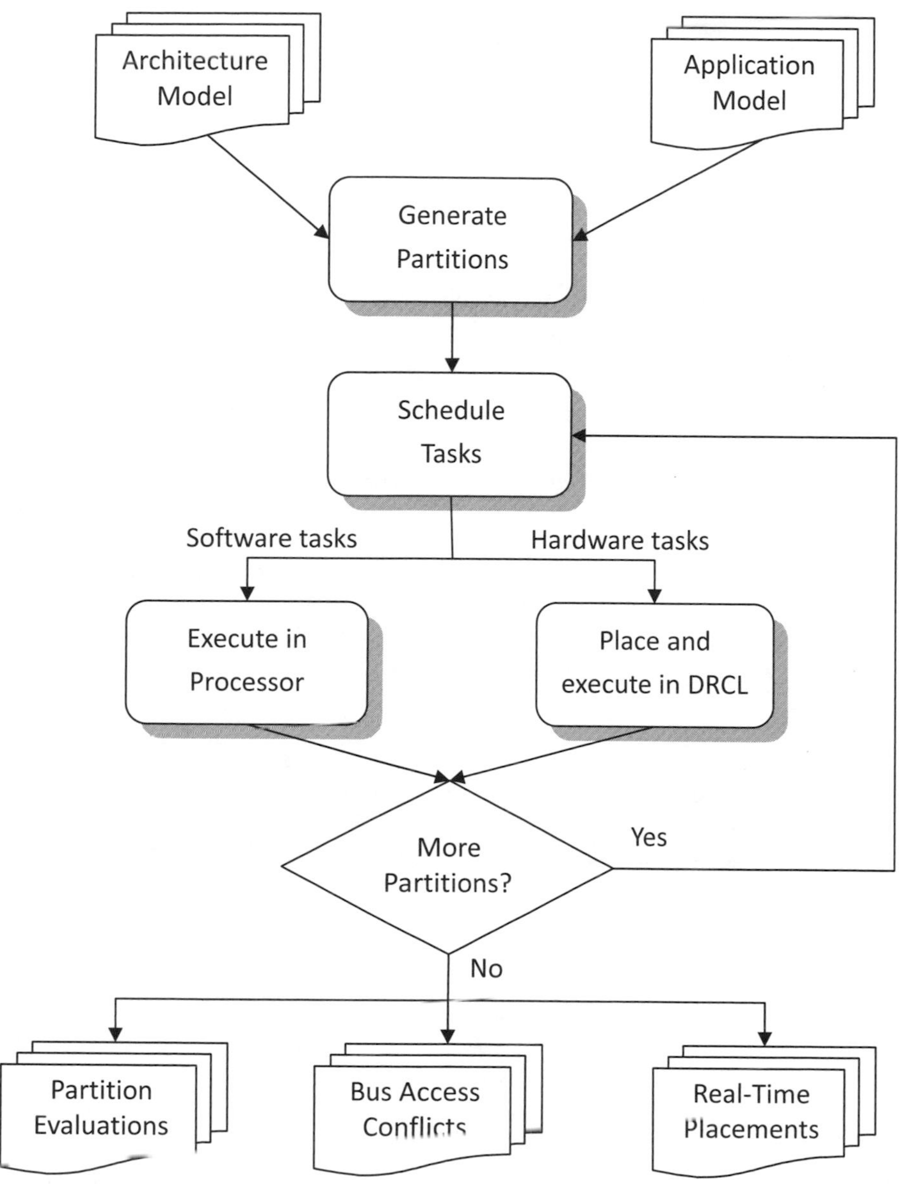

Figure 6.2: Perfecto simulation flow.

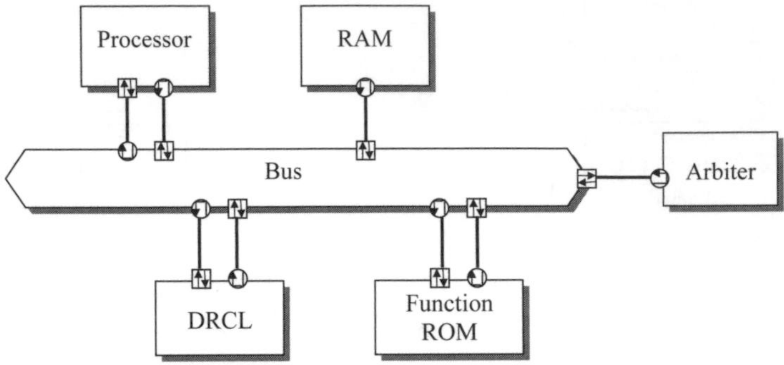

Figure 6.3: Dynamically reconfigurable system architecture model in SystemC.

guage [142] was used to develop this architecture model because design space exploration is being performed at the system level and because the design contains both hardware and software functions. In this architecture, a simple bus model is used as the communication infrastructure for hardware tasks. Here, "simple" means that there is no pipeline and no split transaction. The bus model eliminates the need to do global routing after tasks are placed into the DRCL. Function ROM is a memory storage to save configurations (e.g., bitstreams) that will be loaded into and executed in DRCL. The other system models are described in the rest of this subsection.

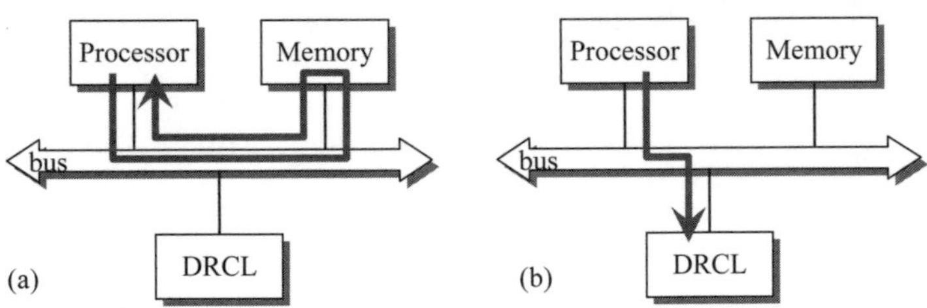

Figure 6.4: Behavior of processor.

6.4.1.1 Processor Model

A processor is required to execute the software in a dynamically reconfigurable system. Besides controlling peripheral devices, it has two main behaviors. One is using the bus to access data from memory for executing software instructions, as shown in Figure 6.4(a). The other behavior is to issue commands to DRCL, such as reconfigure or execute, as shown in Figure 6.4(b). These two behaviors of the processor are simulated at the transaction level.

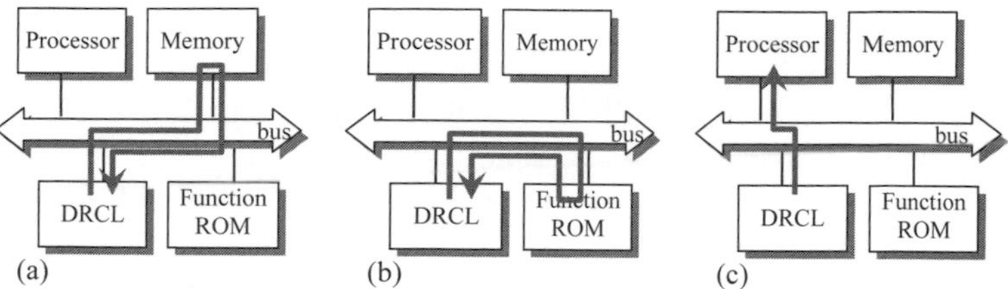

Figure 6.5: DRCL behaviors.

6.4.1.2 Bus Arbiter Model

An arbiter model is required as there is more than one master device on the bus in a dynamically reconfigurable system, including the processor and the DRCL. The main behavior of an arbiter is to arbitrate when more than one request is made for the bus. The arbiter selects the most suitable request to grant bus access according to the following policy, while the other requests are kept in the waiting queue.

1. If the current request is a locked burst request, then it is always selected.

2. If the last request had its lock flag set and the corresponding master is "requesting" again, this request is selected from the waiting queue and returned.

3. The request with the highest priority (smallest number) is returned from the waiting queue.

6.4.1.3 DRCL Model

There are three main behaviors in a DRCL model. The first is memory access. As shown in Figure 6.5(a), when a DRCL receives an execute command from the processor, the DRCL will access memory according to the address parameters. Next, DRCL may request a bitstream from the function ROM as shown in Figure 6.5(b). As shown in Figure 6.5(c), DRCL may issue a response command to the processor when it completes a task. Note that the configuration controller is embedded within the DRCL model, which is similar to the embedded ICAP controller in Xilinx Virtex FPGAs.

Reconfigurable logics can be divided into two configuration styles, namely full-configure (static) and partial-configure (dynamic). The DRCL model in Perfecto can simulate both types of configurations. However, SystemC does not allow dynamic binding of modules with their behaviors and also does not allow a module to have multiple behaviors that can be configured at run time. We can thus say that SystemC does not support reconfiguration of any kind. There are several workarounds for simulating reconfiguration in SystemC. A straightforward method is the static binding of a SystemC module to multiple behaviors and then selecting one of the behaviors for dynamic execution. This method is the simplest, but quite inflexible because for every new function, we need to modify the DRCL model. Similar to DRCF [147] and DRC [151], Perfecto adopts this method because it is simple and can be simulated quickly. For design space exploration, the speed of simulation is of utmost importance. Other methods include the use of C function pointers [162] and C++ templates, all of which might cause an overhead in SystemC simulation performance and are thus not very suitable for design space exploration. Further, multiple `sc_threads` are used in the DRCL for modeling partial reconfiguration. To avoid the use of function pointers, Perfecto uses a function table and a task table, with interfaces for automatic insertion of new functions and new tasks.

6.4.1.4 Architecture Model and Parameters

In Perfecto, the basic architecture model as illustrated in Figure 6.3 is simulated using the above described models of the processor, the memories, the arbiter, and the DRCL. A software task executes in the processor model by accessing the memory. A reconfigurable hardware task executes in the DRCL model by accessing the memory for input and output data. Communications between a software task and a hardware task are accomplished by the processor and the DRCL models. A hardware function reconfiguration is accomplished by the DRCL model by accessing the function ROM. The arbiter grants access to the bus for memory accesses by the processor and the DRCL.

If Perfecto simulates only a fixed basic architecture model, then it will be of little use for a designer who wants to experiment with different system design alternatives. Thus, Perfecto allows a system designer to tune the basic architecture model through several architecture parameters, as described in Definition 6.1.

DEFINITION 6.1 *An architecture model is defined as a tuple* $S = \langle W_{bus}, N_{mem}, T_{mem}, N_{slice}, A_{part}, A_{sched}, A_{place} \rangle$, *where*

- W_{bus} *is the bus width. The basic unit is a word of 4 bytes. A designer may specify the bus width in units of word. This parameter affects the memory access counts of an application running in the target system.*

- N_{mem} *is the memory size. This is a multiple of 4 because the smallest memory unit is 4 bytes. The parameter affects the application response time.*

- T_{mem} *is the memory access time. This parameter is the time for one memory access. This parameter also affects the application response time.*

- N_{slice} *is the DRCL size. This is the total number of slices in a DRCL.*
- A_{part} *is the partitioning algorithm. The default partitioning algorithm is function-based partitioning. Users can implement their own partitioner for task mapping or choose an optional method, including the common-hardware-first or the random partitioning as described in Section 6.4.3.*
- A_{sched} *is the scheduling algorithm. The default scheduling algorithm is simply FIFO. Users can implement their own scheduler to optimize an application. Another scheduler also currently implemented in Perfecto is an energy-efficient hardware-software co-scheduler [126] for reconfigurable systems.*
- A_{place} *is the placement algorithm. The default placement algorithm is a rule-based one, which is described in Section 6.4.5. Users can implement their own placer to place the hardware tasks into DRCL.* □

As an illustrative example, the architecture model could be as follows. The bus width is 2 words, the memory size is 100 MB, the memory access time is 10 ns, the maximum DRCL slices available is 5, and the partitioning, scheduling, and placement algorithms are all the default ones.

6.4.2 Application Model

Besides using parameters to model a user-desired, dynamically reconfigurable system architecture for simulation, Perfecto further allows designers to specify the application model that represents an application to be executed on the reconfigurable architecture model. An application is defined as a set of concurrent tasks with possible precedence relations among the tasks. Thus, an application can be formalized as follows.

DEFINITION 6.2 *An application is represented by a directed acyclic graph* $G(V, E)$, *where*

- V *is a set of nodes representing the application tasks. Each task* T_i *invokes a function* F_j, *which is represented by the tuple* $F_j = (f_{name}, t_{sw}, t_{cfg}, t_{hw}, n_{slice}, f_{code})$, *where*
  - f_{name} *is a unique function name, such as DES, AES, DCT, etc.*
  - t_{sw} *is the computation time of the software implementation of the function, without considering memory access time*
  - t_{cfg} *is the configuration time of the hardware implementation of the function*
  - t_{hw} *is the computation time of the hardware implementation of the function*

- n_{slice} *is the area of DRCL required by the function, in terms of the number of slices, where a slice is basic unit of configuration such as a frame, column, or tile in Xilinx Virtex FPGAs*
- f_{code} *is the function behavior code, implemented as SystemC transaction-level code and is used to model the function behavior; see Figure 6.6 for an example*

Note that the same function can be invoked by different tasks, but without any data sharing between the different invocations.

- *E is a set of edges representing the task precedence relations. An edge* $(u, v) \in E$ *means that task v must wait for task u to complete before starting execution.* □

An application is specified by a designer through several task parameters extracted from Definition 6.2, which includes the set of tasks and functions, the mapping between the tasks and the functions, the six function attributes, and the precedence relations among the tasks. Note that modeling a new application into an appropriate set of tasks could be a complicated job, which is outside the scope of this work.

To illustrate the above-discussed task parameters, we will use a simple application that has 6 tasks invoking 4 functions as given in Table 6.1(a), where the mappings between tasks and functions are given. It is also specified that task T3 starts execution only after task T5 is done. The function attributes specified by the user are shown in Table 6.1(b). For example, function F3, when implemented in software, requires 1300 ns execution time without considering memory accesses and when implemented in hardware requires 150 ns configuration time, 600 ns execution time, and uses 2 slices. A generic example of function behavior code is shown in Figure 6.6. Thus, Table 6.1 and Figure 6.6 depict the task parameters.

The architecture parameters for this example are similar to what were specified previously at the end of Section 6.4.1.4. The results of performance evaluation for this simple example using Perfecto are presented in Section 6.4.7.

Sometimes a designer might want to evaluate a specific reconfigurable system architecture along with some specific combinations of partitioning, scheduling, and placement algorithms; however, he/she might not want to target it for some specific application. To perform such application-independent evaluations, we designed a randomized application model interface of Perfecto based on the *Task Graphs For Free* (TGFF) tool [54], as described in Section 6.4.2.1.

6.4.2.1 TGFF Interface

Task Graphs For Free (TGFF) [54] is a tool for generating random task graphs based on some user-specified requirements on the graphs. TGFF has been widely used by many academic and industrial tools for computer-aided design.

Table 6.1: Tasks in Illustration Example
$V = \{$T1, T2, T3, T4, T5, T6$\}, E = \{(T5, T3)\}$

Tasks (V)	T1	T2	T3	T4	T5	T6
Function	F1	F3	F2	F2	F4	F3

(a) Task Graph $G(V, E)$

f_{name}	$t_{sw}(ns)$	$t_{cfg}(ns)$	$t_{hw}(ns)$	n_{slice}
F1	200	0	0	0
F2	1000	100	500	1
F3	1300	150	600	2
F4	2000	200	1000	1

(b) Function Parameters

Since Perfecto is used for design space exploration, an interface to TGFF eliminated the need for users to specify an exact application for evaluation. The TGFF interface allows users to thoroughly evaluate a reconfigurable system along with its partitioning, scheduling, and placement algorithms because the task graphs are randomly generated.

TGFF generates the task set information from a template by varying the seed for the random number generator per template. The template parameters to be defined include the number of task graphs, the average number of functions in a task graph, and the function attributes. The textual representation of task graphs generated by TGFF is then automatically parsed by Perfecto into intermediate task data structures that can then be used for partitioning, scheduling, and placement, thus automating design space exploration and performance evaluation.

6.4.3 Partitioning

A partitioning algorithm maps each task in an application model to either software or hardware based on some estimation criteria such that the task is executed either on the microprocessor or in the DRCL, respectively. Formally, it is defined as follows.

DEFINITION 6.3 *Given an architecture model S and an application model $G(V, E)$, the partitioning algorithm is defined as $A_{part}(G(V, E), S) = \mu(V, \{0, 1\})$, where $\mu(T_i) = 0$ represents that T_i is mapped to software, and $\mu(T_i) = 1$ represents that T_i is mapped to hardware.* □

By selecting the mapping criteria, we can have different partitioning algorithms. A mapping result of a partitioning algorithm is called a *system*

```
bool func_rom::func_1(unsigned int priority, int task_no, int CT,
int ET,
    unsigned int src1, unsigned int src2, unsigned int src3, unsigned int des1,
    unsigned int des2, int slice_no, int sw_hw) {
  double time_1 = 0;
  int time_2 = 0;
  int data1[src3], data2[src3];

  time_1 = sc_simulation_time(); // record start time
  if(CT != 0){ time_2 = CT; wait(CT, SC_NS); } // CT: configuration time
  func_rom_bus_port->burst_read(priority, task_no, data1, src1, src3);
  func_rom_bus_port->burst_read(priority, task_no, data2, src2, src3);
  for(unsigned int i=0; i<src3; i++)    data1[i] = data1[i] + data2[i] + 1;
  wait(ET, SC_NS); // ET: execution time, without communication
  func_rom_bus_port->burst_write(priority, task_no, data1, des1, des2);
  time_1 = sc_simulation_time() - time_1; // calculate total time
  func_rom_bus_port->direct_response(sw_hw, task_no, (int)time_1-ET-CT,
                                     time_2, slice_no);
  return true;
}
```

Figure 6.6: SystemC function behavior code for simple example.

partition or simply *partition*. A partitioning algorithm thus generates a set of system partitions. For experiment purposes, Perfecto implemented three hardware-software partitioning algorithms, namely function-based, common-hardware-first, and random partitioning, as described in the following.

The *function-based partitioning* algorithm maps each function into hardware and/or software according to the function attributes. For a set F of functions, at most $2^{|F|}$ partitions are generated, irrespective of the number of tasks. Hence, in a partition, even if the same function is invoked by multiple tasks, all of them are mapped to the same implementation, either hardware or software depending on the partition. Though non-exhaustive, this mapping greatly reduces the number of partitions generated since the number of functions is usually smaller than the number of tasks.

The *common-hardware-first partitioning* algorithm first counts the number of task graphs each function is invoked, denoted by $c(f)$ for a function $f \in F$. If $c(f) > 1$, then f is called a *common* function. The common functions are then sorted in descending order according to $c(f)$. Given a parameter $k > 0$ representing the number of common hardware functions desired, we map the first k common functions from the ordered list into hardware and map the rest of the functions into software. This partitioning algorithm is useful because several scheduling and placement algorithms often employ heuristics based on common hardware functions, such as the energy-efficient hardware-software scheduling [91] and the configuration-reuse scheduling and placement methods [137, 155].

A third random partitioning method generated random partitions according to user requirements. Perfecto did not employ any complex partitioning algorithm because the purpose was not to propose a new partitioning algorithm; the purpose was merely to check if the framework can be used to efficiently evaluate dynamically reconfigurable system designs.

Though several partitioning algorithms were implemented in Perfecto, users have to compare the results of the different partitioning algorithms manually after Perfecto has generated the partitions by applying the algorithms one by one. The partitioning results of different algorithms can be compared by considering either the total number of partitions generated by an algorithm or the quality of the partitions generated. After Perfecto applies the partitioning, scheduling, and placement algorithms, the quality of the partitions can be gauged. Details of the characteristics of partitions can be found in Section 6.4.6. An ideal partitioning algorithm is one that can generate the optimal partition within a minimal number of partition results. Optimality in quality and minimality in quantity are conflicting goals and thus a real partitioning algorithm can only generate a near-optimal partition in a manageable number of partitions. Users can thus select a partitioning algorithm based on the desired trade-off between quality and quantity.

Users can also invent a new partitioning algorithm and implement it into Perfecto to check if the evaluated performance improves. Perfecto was modularly designed such that independent data structures were used for the set of all tasks, the set of software tasks, and the set of hardware tasks. Thus, a user merely has to re-write the `Generate_Partition()` function to implement a new partitioning algorithm. Well-defined interfaces between this function and the consequent scheduling algorithm have thus helped users to implement new algorithms into Perfecto.

6.4.4 Scheduling

A scheduling algorithm associates an ordering to the set of software tasks that are ready so that they will be executed in that order on the microprocessor. It also associates an ordering to the set of hardware tasks that are ready so that they will be placed and executed in that order in the DRCL. This is defined formally in Definition 6.4.

DEFINITION 6.4 *Given a set of tasks V, a system partition μ, and a time instant t, a scheduling algorithm is defined as $A_{sched}(V, \mu, t) = \langle \gamma_{sw}(V, t), \gamma_{hw}(V, t) \rangle$, where γ_{sw} associates a strict total order to the set of software tasks that are ready at time t, i.e., $\{T_j \mid \mu(T_j) = 0, (T_i, T_j) \in E \rightarrow T_i \text{ has terminated}\}$ and γ_{hw} associates a strict total order to the set of hardware tasks that are ready at t, i.e., $\{T_j \mid \mu(T_j) = 1, (T_i, T_j) \in E \rightarrow T_i \text{ has terminated}\}$.* □

By selecting the ordering criteria, different scheduling algorithms can be designed. A specific order of tasks generated by a scheduling algorithm is called a *schedule*, and the time required to complete the tasks in that order is called the *schedule length.* Two scheduling methods were implemented in Perfecto, namely a default FIFO method and an energy-efficient hardware-software co-scheduling algorithm [91].

Given the same system partition μ, different scheduling algorithms result in different schedule lengths. Smaller schedule lengths are desired so that the set of tasks complete execution sooner. The overall system schedule length is the maximum of the hardware schedule length and the software schedule length. However, due to task dependencies the hardware schedule and the software schedule usually affect each other, thus in general it might not be possible to optimize only one of the two schedules without considering the other one. Perfecto allows the application of different scheduling algorithms on the same set of partitions. Users can select the scheduling algorithm that results in the smallest system schedule length. However, it must be noted here that the hardware schedule length is directly affected by the placement algorithm.

The intent here was not to propose a new scheduling algorithm. Users can always implement an existing algorithm such as the Horizon or Stuffing algorithms [178], classified stuffing [88], and real-time relocatable task scheduling [43], or invent a new one. Since the algorithms themselves are outside the scope of this work, interested readers should refer to [91] for further details on the energy-efficient hardware-software co-scheduling algorithm that was implemented in Perfecto. Integrating new scheduling algorithms into Perfecto also allows another dimension of exploration, whereby we can tune a scheduling algorithm or select the best scheduling algorithm for a specific application. Due to modular design and the well-defined interfaces in Perfecto, users merely have to modify or rewrite the `Scheduler()` function that takes two sets of tasks Q_1 and Q_2 and sorts them according to some criteria, where Q_1 is a set of ready hardware tasks and Q_2 is a set of ready software tasks.

6.4.5 Placement

A placement algorithm tries to find a feasible location in a fixed-size DRCL for a given hardware task of some fixed size. The placement algorithm depends on the underlying configuration model, namely paged 1-dimensional, segmented 1-dimensional, or 2-dimensional, where the basic units of configuration are a fixed-size *slot*, a *column*, or a *tile*, respectively. In Perfecto, an abstract model is used where the basic unit of configuration is simply called a *slice*. Thus, in Perfecto the underlying configuration model could be any of the three.

DEFINITION 6.5 *Given a DRCL of N_{slice} slices, a list L_{used} of spaces allocated to tasks, a list L_{free} of free spaces, and a task T_j of size $n_{slice}(T_j)$ slices to be placed, a placement algorithm is defined as $A_{place}(T_j, L_{used}, L_{free}) = \langle loc, L'_{used}, L'_{free}\rangle$, where loc is either NULL or a pointer to a feasible location for the task T_j in the DRCL such that $n_{slice}(loc) \geq n_{slice}(T_j)$. Suppose after placing T_j in loc, the used part of loc is denoted as loc$'$ and the remaining free*

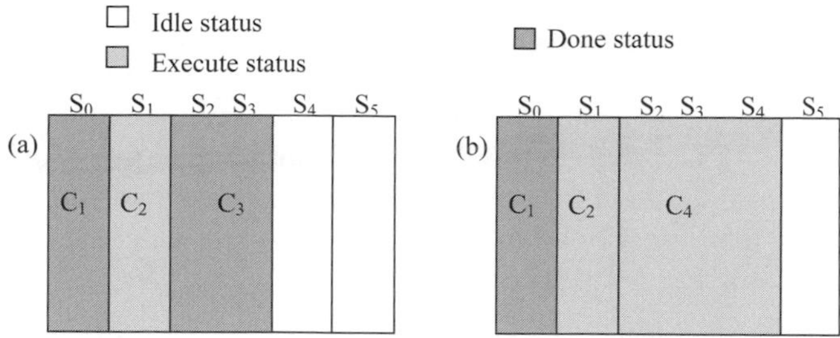

Figure 6.7: Hardware task placement in Perfecto.

part, if any, is denoted by loc″, i.e., $n_{slice}(loc) = n_{slice}(loc') + n_{slice}(loc'')$, *where* $n_{slice}(loc') = n_{slice}(T_j)$. *Then, the lists are updated as follows.* $L'_{used} = L_{used} \cup \{loc'\}$. $L'_{free} = Merge\_Adj(L_{free} \backslash \{loc\} \cup \{loc''\})$, *where the function Merge_Adj() tries to merge adjacent free spaces into contiguous blocks of free space.* □

By changing the selection criteria for the feasible location to place a task, different placement algorithms can be invented. Currently in Perfecto, the following placement policy is used. The DRCL selects a block for configuration according to the following rules:

1. If there exists a block that is already configured with the same circuit, but not executing currently, i.e., it is in the *done* status, then reuse the block by selecting it for the current task.

2. If there exists a configured block with the same slice count, but with a different function and at the *done* status, then configure the new task in this block.

3. If there exists an unconfigured block, i.e., in the *idle* status, with enough slices, then configure in that block.

4. If there is no free block with enough slices, the blocks at the *done* status will be released into the *idle* status, then check rule 1 to rule 3 again.

An example is shown in Figure 6.7(a), where a DRCL is divided into several slices, S_0 to S_5. Suppose there are three circuits already configured in the DRCL; however, C_1 and C_3 are at the *done* status, and C_2 is in the *execute* status. If a request for circuit C_2 is issued to the DRCL, according to the above placement rules, this task will be configured and executed in S_0. If a request is received for circuit C_4 by the DRCL, and if circuit C_4 needs 3 slices, then circuit C_3 will be released first, and then circuit C_4 configured into S_2 to S_4, as shown in Figure 6.7(b).

The above described is only a sample 1-dimensional placement strategy implemented in Perfecto. The algorithms themselves are not the focus of this framework. Users can always implement 2-dimensional and other placement strategies such as the multi-objective placement [124] and integrate them into Perfecto for further evaluation of both algorithms and systems. The function `Placer()` takes a task, a list of used blocks, and a list of free blocks. It then computes the ideal location for the task and updates the two lists as described in this section. The users have to simply modify or design a new function to replace the default Perfecto `Placer()`.

As a final note, in a real system, a bitstream must be *relocated* to the target location before it can be used for configuration, where relocation simply means a change of the major address (MJA) to the target column or tile. This relocation can be achieved using software such as Parbit or using hardware filters such as REPLICA. Since the bitstream relocation does not affect the performance of the system significantly, it is abstracted in Perfecto.

6.4.6 Performance Evaluation Results

Applying the partitioning, scheduling, and placement algorithms described in Sections 6.4.3, 6.4.4, and 6.4.5, respectively, to a user-parameterized reconfigurable architecture model and a user-specified application model, the built-in simulation capabilities of SystemC are used to simulate the system. During simulation, as shown in Figure 6.2, several performance readings are collected and three results are generated, including partition evaluations, bus access conflicts, and real-time placements.

Before describing how the results are evaluated by Perfecto, an understanding of some basic terminologies and definitions is required. Given a task t that invokes a function $F = (f, t_{sw}, t_{cfg}, t_{hw}, n_{slice}, f_{code})$ in a partition P, we use $\lambda(t, P) = (f, u)$ to denote the implementation of the task t (i.e., function f) as a software function if $u = 0$ and as a hardware function if $u = 1$. The computation time $ET(t, P)$, the configuration time $CT(t, P)$, and the reconfigurable resource requirements $RR(t, P)$ are defined as follows. For a software task, $ET(t, P) = t_{sw}$, $CT(t, P) = 0$, and $RR(t, P) = 0$, while for a hardware task $ET(t, P) = t_{hw}$, $CT(t, P) = t_{cfg}$, and $RR(t, P) = n_{slice}$.

For each task t in each system partition P, Perfecto accurately evaluates the total *Task Execution Time* ($TET(t, P)$), which is the sum of the computation time ($ET(t, P)$), the configuration time ($CT(t, P)$), the memory access time ($MAT(t, P)$), and the bus wait time ($BWT(t, P)$). The first two are as defined above, and the last two are obtained through simulation.

For each partition P, Perfecto evaluates five attributes of the partition called *partition evaluations*, which include the total *Partition Execution Time* (PET) in nanoseconds, the *Average DRCL Utilization* (ADU %), the *maximum number of DRCL slices used* (MS), the percentage of *Average Configuration Time* (ACT %), and the percentage of *Average Bus Waiting Time* (AWT %). Out of the five attributes, the values of PET and MS depend on the scheduler and the placer chosen in Perfecto, respectively. The other three attributes are defined as follows.

$$ADU(P) = \frac{\sum_t (TET(t,P) \times RR(t,P))}{PET \times MS} \tag{6.1}$$

$$ACT(P) = \frac{\sum_t CT(t,P)}{\sum_t TET(t,P)} \tag{6.2}$$

$$AWT(P) = \frac{\sum_t BWT(t,P)}{\sum_t TET(t,P)} \tag{6.3}$$

The bus access conflicts show the real-time information of the number of tasks competing for bus access and also the tasks that are actually making requests. From this information, a designer can detect if there is a bottleneck in system performance. The real-time placement information for each task in each partition can be used for further tuning and optimizations.

After simulation, by analyzing the above results generated by Perfecto, a designer can then decide to select one or more partitions that best fits his/her needs. The criterion could be the least total execution time, or the least average DRCL utilization, or the one with the least average bus waiting time. All of these results would be more apparent and intuitive through application examples, as described in Section 6.4.7.

6.4.7 Application Examples

The implementation of Perfecto was in IEEE 1666-2005 SystemC on a Linux Fedora Core 3 workstation with Intel Pentium 4 2.4 GHz CPU and 1 GB RAM. Perfecto was applied to several designs. We use a simple example to illustrate the framework and then show its application to a more complex real-world network security example. Note that the partitioning, scheduling, and placement algorithms applied to the examples in this section are all the default ones in Perfecto such that we can focus on the framework itself.

The simple illustration example was introduced in Section 6.4.2, which has 6 tasks invoking 4 functions, with a precedence relation (T5, T3). Note that from the function attributes in Table 6.1, we can conclude that F1 has a software implementation only. Perfecto uses function-based partitioning to generate the partitions for the example as shown in Table 6.2. Partition P0 has the most hardware tasks and P7 is the all-software partition. In the following, we show how Perfecto helps a system designer make intelligent partition decisions, optimize system performance, and evaluate task placement strategies.

6.4.7.1 Making Intelligent Partition Decisions

All the eight hardware-software partitions for this example were evaluated by Perfecto as shown in Table 6.3, which gives the composition of time and the number of slices required by each task. Take partition P0 as an example. Its total execution time is 2033 ns, the average DRCL utilization is 54.63%,

Table 6.2: Partitions for Simple Illustration Example

Partition#	Function Name (f_{name})				Num Tasks	
	F1 {T1}	F2 {T3, T4}	F3 {T2, T6}	F4 {T5}	SW	HW
P0	0	1	1	1	1	5
P1	0	1	1	0	2	4
P2	0	1	0	1	3	3
P3	0	1	0	0	4	2
P4	0	0	1	1	3	3
P5	0	0	1	0	4	2
P6	0	0	0	1	5	1
P7	0	0	0	0	6	0

0: Implement in Software, 1: Implement in Hardware,
SW: Number of software tasks, HW: Number of hardware tasks

the maximum usage of DRCL is 3 slices, the average configuration time is 11.96%, and the average bus waiting time is 2.28%. Further, the execution time of task T4 in partition P0 is 752 ns, which includes pure execution time (500 ns), configuration time (100 ns), memory access time (90 ns), and bus waiting time (62 ns). From Table 6.3, some useful conclusions can be drawn as follows, which will help a reconfigurable system designer make intelligent design choices.

1. The most-hardware partition P0 requires the least execution time, while the all-software partition P7 requires the most execution time. This observation, though intuitively expected, might not be true if the reconfiguration time is very large.

2. The execution times for partitions P1 and P2 are quite close; however, P2 uses fewer DRCL slices than P1. If P2 is selected for system design, we can use a smaller DRCL (2 slices instead of 5) for this application. Architecture exploration can thus be performed.

3. Partition P4 gives a very good trade-off between hardware and software because its total execution time (2623 ns) is quite close to that of the most-hardware partition P0 (2033 ns) and its average bus waiting time (0.2%) is negligible just like that of the all-software partition P7, which instead has a much higher execution time (7351 ns).

6.4.7.2 Optimizing Performance

Perfecto not only helps designers make intelligent hardware-software partitioning choices but also makes debugging performance bottlenecks easier by

Table 6.3: Evaluation by Perfecto of Hardware-Software Partitions for Simple Example

$P\#$	PET	ADU	MS	ACT	AWT	$\frac{CT \mid MAT \mid BWT}{\text{Task Execution Time}}$ (ns), where Task Execution Time $= ET + CT + MAT + BWT$																	
	(ns)	(%)		(%)	(%)	T1			T2			T3			T4			T5			T6		
P0	2033	54.63	3	11.96	2.28	0	240	0	150	20	3	100	90	0	100	90	62	200	60	40	0	20	0
							440			773			690			752			1325			620	
P1	3119	29.54	5	7.27	2.80	0	240	0	150	20	4	0	90	0	100	90	61	0	60	0	150	20	89
							440			774			590			751			2085			859	
P2	3083	17.16	2	4.62	1.23	0	240	0	0	20	0	0	90	0	100	90	51	200	60	29	0	20	0
							440			1320			590			741			2085			1320	
P3	5169	5.19	1	1.54	0.94	0	240	0	0	20	0	0	90	9	100	90	52	0	60	0	0	20	0
							440			1320			599			742			2085			1320	
P4	2623	33.45	5	9.16	0.20	0	240	0	150	20	2	0	90	0	0	90	0	200	60	0	150	20	9
							440			772			1090			1090			1285			779	
P5	4709	13.20	4	4.79	0.22	0	240	0	150	20	3	0	90	0	0	90	0	0	60	0	150	20	11
							440			773			1090			1090			2085			781	
P6	5265	4.88	1	3.06	0.00	0	240	0	0	20	0	0	90	0	0	90	0	200	60	0	0	20	0
							440			1320			1090			1090			1285			1320	
P7	7351	0.00	0	0.00	0.00	0	240	0	0	20	0	0	90	0	0	90	0	0	60	0	0	20	0
							440			1320			1090			1090			2085			1320	

PET: Partition Execution Time, ADU: Average DRCL Utilization
MS: Maximum Usage of DRCL Slices (out of totally 5 slices), ACT: Average Configuration Time
AWT: Average Bus Waiting Time, ET: Pure Execution Time $\in \{t_{sw}, t_{hw}\}$, CT: Configuration Time $\in \{0, t_{cfg}\}$,
MAT: Memory Access Time, BWT: Bus Wait Time

providing designers with detailed real-time information on how and when the tasks in each partition compete for bus access. From Table 6.3, we can observe that task T6 in partition P1 has the largest bus waiting time of 89 ns among all the tasks in all partitions. Further, this also directly reflects on the average bus waiting time of partition P1 (2.8 ns), which is largest among all the 8 partitions. To debug this performance bottleneck, we can analyze the bus conflict accesses reported by Perfecto. Figure 6.8 shows the aggregate and the individual task bus accesses for Partition P1 along the time axis. Analyzing the aggregate diagram, we can immediately identify there is a performance bottleneck in partition P1, where the maximum number of concurrent bus accesses is 3. Suppose the system designer would like to solve this bottleneck. Through the individual task accesses provided by Perfecto, we can identify the tasks that are competing in this bottleneck of 3 concurrent bus accesses, namely tasks T1, T3, and T6. We can also exactly pinpoint the time (200 ns) during which the three tasks are competing for the bus. These information details help designers resolve the bottlenecks and optimize their designs.

6.4.7.3 Evaluating Task Placements

Besides guiding system designers in making intelligent partitioning decisions and in resolving performance bottlenecks, Perfecto also helps designers tune hardware task placement algorithms by providing designers with detailed real-time placement information for each task in the DRCL area. In the simple illustration example from Table 6.3, we can observe that partition P1 uses maximum 5 DRCL slices, but has quite a low average DRCL utilization of 29.54% only. Thus, we would like to investigate the task placement in P1, which is shown in Figure 6.9, for a DRCL of 5 slices. The numbers in the top-left corner of each block represent the time when the blocks were placed. For example, task T2 was placed and configured at 6 ns, T4 at 8 ns, T6 at 10 ns, and T3 at 2528 ns. The status of a slice indicates if it is idle, executing, or done. A slice in an idle or done state can be reused for placing newly arrived tasks. As we can observe there are large periods of time, and several slices are unused. Intuitively, we could see that if we delay the execution of T6 to 780 ns, we would place it in slices S0 and S1 following T2. Further, we could also delay T4 and place it in S1, following T6. Finally, we would place T3 also in S1 following T4. Thus, only 2 slices are enough, without affecting the total execution time of P1. Note that T3 can start only at 2,528 ns due to its precedence constraint with T5, a software task in P1. This is an illustration of how the placement information can help designers change the placement algorithm or tune it manually.

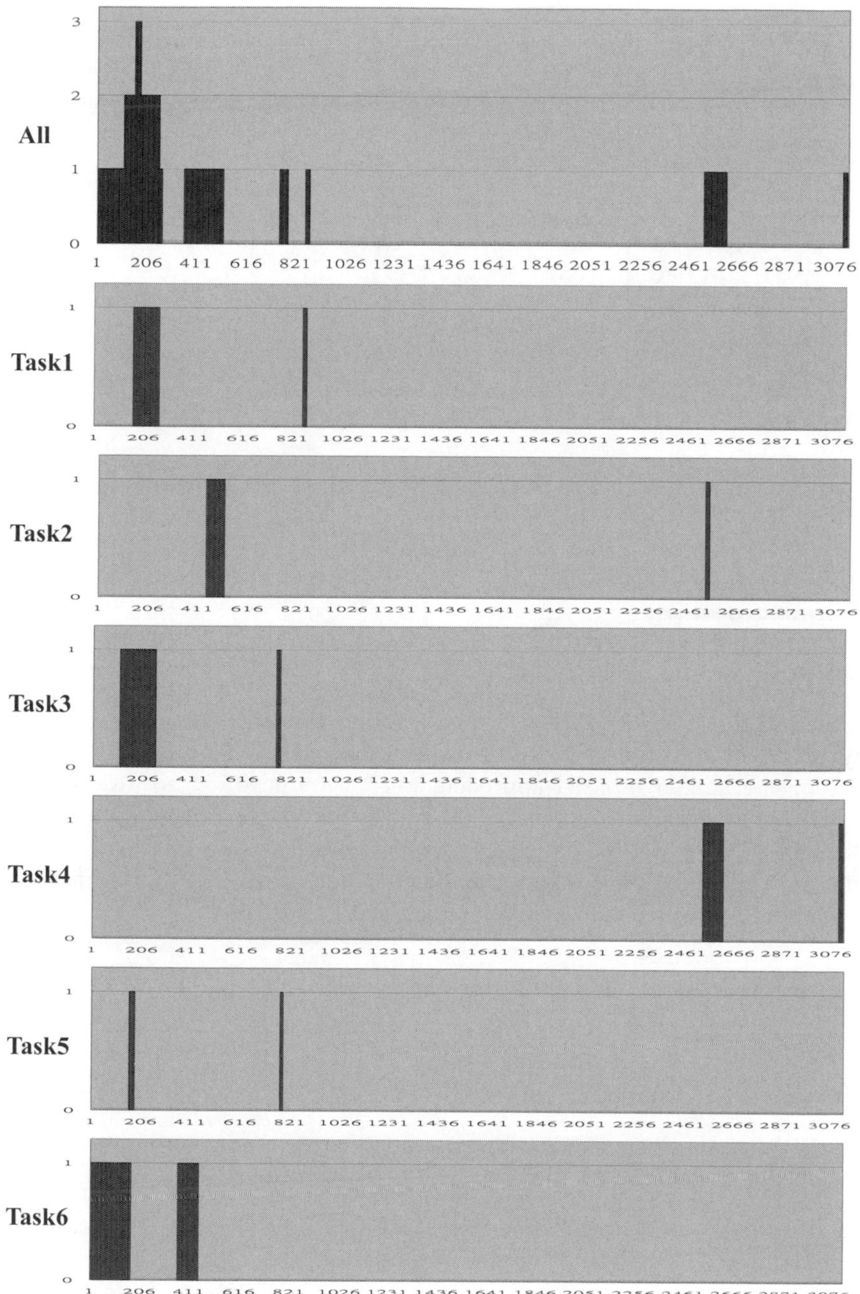

Figure 6.8: Bus access conflicts for partition P1 of simple example.

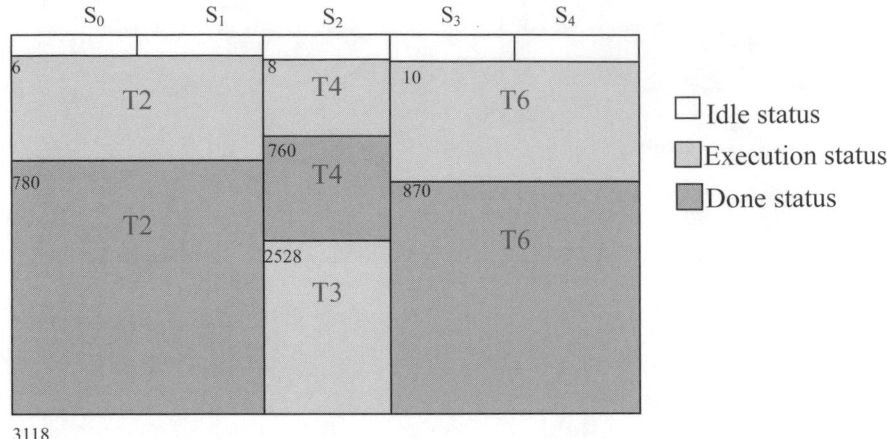

Figure 6.9: Hardware task placement for Partition P1 of simple example.

Summary

We discussed the state of the art in the verification technology for reconfigurable systems. System-level verification techniques, including a formal verification and language-based approach, were introduced. Hardware-software coverification techniques, including cosimulation and prototyping, were also discussed. Finally, we introduced a SystemC-based design space exploration framework called Perfecto that can be used for evaluating performance of reconfigurable systems and also for designing OS4RS.

Exercises and Problems

1. What makes the verification of reconfigurable systems more complex than conventional systems?

2. How can semi-formal verification techniques be employed for reconfigurable systems? (Hint: semi-formal verification means the integration of formal verification with dynamic verification such as model checking and simulation.)

3. Discuss the pros and cons of using a formal language in reasoning about reconfigurable systems.

4. Compare the advantages and disadvantages in using formal verification techniques versus simulation techniques for dynamically reconfigurable systems.

5. Describe how Perfecto differs from conventional verification frameworks for dynamically reconfigurable systems.

6. How can a user take full advantage of Perfecto when designing reconfigurable systems?

7. Compare Perfecto with SyCERS in terms of reconfigurable system design and verification.

8. **Project Theme**: Create a system-level framework for analyzing some features of reconfigurable systems. For example, you could use the SystemC language to model reconfigurable components and use SystemC built-in simulation capabilities to evaluate the performance of reconfigurable components when the number of partially reconfigurable regions (PRR) is fixed.

7

Dynamically Partially Reconfigurable System Design Implementation

With rapid technology progress, FPGAs are getting more and more powerful and flexible in contrast to inflexible ASICs. FPGAs, such as Xilinx Virtex II/II Pro, Virtex 4, and Virtex 5, can now be partially reconfigured at run-time for achieving higher system performance. Partial reconfiguration means that one part of the FPGA can be reconfigured while other parts remain operational without being affected by reconfiguration. An embedded system along with FPGAs, which is usually called a *Dynamically Partially Reconfigurable System* (DPRS), can enable more and more intensive applications to be accelerated in hardware at run-time, and thus the overall system execution time can be reduced [95, 213].

There exist many related research works in which the capability for partial reconfiguration is used to enhance the flexibility and the performance of their proposed systems. Implementing a DPRS is a very complex task compared to a traditional embedded system. The DPRS includes not only the traditional software and hardware applications in a traditional embedded system, but also the dynamically partially reconfigurable hardware designs running on the FPGA. However, access to the detailed information on the implementation of DPRS is distributed, without any complete user or reference manual. Hence, in this chapter, we will guide the designer to implement a DPRS so that he or she can avoid spending a lot of time on searching and studying all the related documents and materials.

Section 7.1 introduces the capability for partial reconfiguration on the Xilinx Virtex family FPGAs, and then the physical constraints are described. After introducing the concepts, the latest partial reconfiguration design flow, namely *Early Access Partial Reconfiguration* (EA PR) design flow provided by Xilinx [213], is introduced in Section 7.2. Section 7.3 gives a real case study on the design of a simple LED control design for readers to more concretely understand the implementation steps for a DPRS design. Through the case study, readers can learn how to enhance an embedded system with the capability for partial reconfiguration using the Xilinx Virtex family FPGAs. In contrast to the pure hardware design with the capability for partial reconfiguration described in Section 7.3, the software-controlled approach to partial reconfiguration through a specific configuration controller provided with the Xilinx Virtex family FPGAs, namely *Internal Configuration Access Port* (ICAP), is described in Section 7.4. Different from the standalone DPRS described in Section 7.4, the integration of an operating system with the dynamically partially reconfigurable hardware architecture is introduced in Section 7.5. Such an OS is usually called *Operating System for Reconfigurable Systems* (OS4RS).

7.1 Partial Reconfiguration on Xilinx Virtex Family FPGAs

To date, Xilinx FPGAs have provided the most complete capability for partial reconfiguration within all commercial FPGAS until now. They are basically organized as two-dimensional array architecture, which is composed of CLBs. The partially reconfigurable region in an FPGA must be a rectangular block, namely *Partially Reconfigurable Region* (PRR). According to the number of hardware functions and the required FPGA resources of every hardware function, a designer has to allocate enough FPGA resources for his/her dynamically partially reconfigurable architecture. To implement the reconfigurable hardware functions, namely *Partially Reconfigurable Modules* (PRMs), we need to follow the partial reconfiguration design flow. Several PRMs can be configured into a PRR. As shown in Figure 7.1, two PRRs are planned in the FPGA, and each PRR includes two different PRMs that can be (re)configured into it. Due to physical configuration constraints, only one PRM can be (re)configured into the FPGA at a time. Given two regions PRR1 and PRR2 and four modules PRM1_a, PRM1_b, PRM2_a, PRM2_b, the DPRS can be configured into one of the four different functional combinations as shown in Figure 7.1, where the arrows indicate the reconfigurations required for changing the functional combination. Note that the set of all PRRs is usually called the *dynamic area*, while the remaining part of the system is called the *static area*. Different Xilinx FPGA families have different physical configuration constraints for implementing the dynamic area, that is, the PRRs. We discuss the basic physical configuration constraints for the different Xilinx Virtex FPGAs before going into the details of the DPRS implementation.

For Xilinx Virtex II/II Pro FPGAs, the basic configuration unit is a CLB column that spans the whole column of the FPGA. The size of a PRR in the FPGA must be a multiple of the CLB column. In the former modular-based-design partial reconfiguration flow [95], the PRR size must be a multiple of the *full* CLB column. However, the latest EA PR design flow [213] does not restrict the PRR size to be a multiple of the *full* CLB column. The PRR can be defined a part of the full CLB column, and thus the remaining FPGA resources in the same CLB column can be allocated for use by the static area. Nevertheless, a CLB column can only be allocated to one PRR. As shown

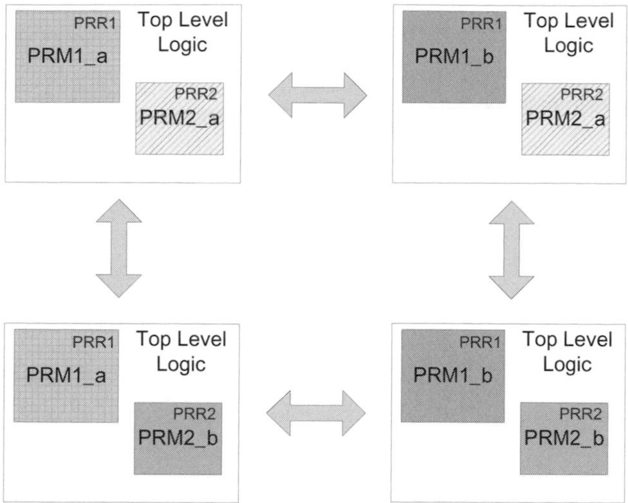

Figure 7.1: Partial reconfiguration on Xilinx Virtex FPGAs.

in 7.2, `PRR_a` and `PRR_b` are correctly placed in the FPGA. `PRR_c` and `PRR_d` overlap with each other in the same CLB column, which results in incorrect placement in the Virtex II/II Pro FPGAs.

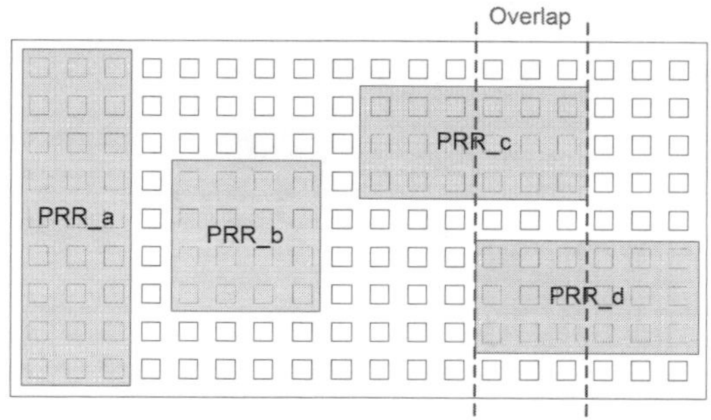

Figure 7.2: PRR allocations on Xilinx Virtex II/II Pro FPGAs.

For Xilinx Virtex 4 FPGAs, the basic configuration unit is a column of 16×1 CLBs that spans the height of a page and not the full chip area, where the FPGA is divided into several regions called pages. Take Virtex 4 XC4VFX12 FPGA, for example, as shown in Figure 7.3. The Virtex 4 FPGA can be divided into eight pages, and a PRR can be placed across one or more pages, such as `PRR_b`. A PRR cannot include any *Input/Ouput Block* (IOB) column, which is in the middle of the Virtex 4 FPGA, because a PRM cannot directly use the I/O blocks of the FPGA to interact with the external I/O devices. Thus, the placement of PRR_d is incorrect. Moreover, similar to the Virtex II/II Pro FPGAs a basic configuration unit can be allocated to only one PRR, so the placement of PRR_e and PRR_f is incorrect. The Xilinx Virtex 5 FPGAs are also divided into a few pages for the placement of PRRs except that the basic configuration unit is 20×1 CLBs.

After understanding the placement constraints for different FPGA families, the allocation for every PRR on the FPGA must include enough FPGA resources to (re)configure its PRMs. Here, the FPGA resource usage can be evaluated by using the reports generated from a synthesis tool and a design analysis tool, such as the Xilinx PlanAhead tool. In the next section, a partial reconfiguration design flow and a specific hardware component, namely bus macros provided by Xilinx, will be described in detail for designing a DPRS.

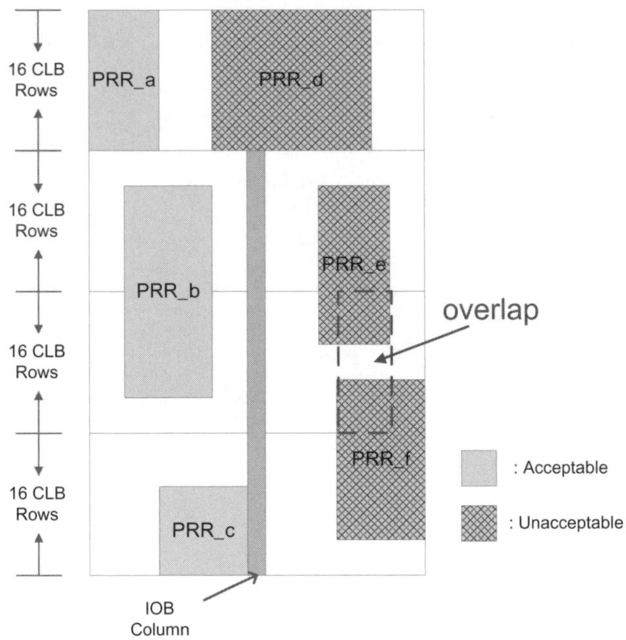

Figure 7.3: PRR allocation on Xilinx Virtex 4 FPGAs.

7.2 Early Access Partial Reconfiguration Design Flow

The *Early Access Partial Reconfiguration* (EA PR) design flow [213] was proposed by Xilinx as a replacement for the former modular design flow [95]. Compared to the modular-based design flow, the EA PR design flow can allow any rectangular size without spanning the whole CLB column for Xilinx Virtex II/II Pro FPGAs. The signals (routes) of the static area can cross through a PRR without the use of bus macros, and thus the timing performance and the implementation of a PRR can be significantly enhanced. The pre-routed CLB bus macros can be used instead of the hard-wired tri-state buffer bus macros. The CLB bus macros can provide more flexibility and greater concentrations of connections between the static and dynamic area [129]. Moreover, the EA PR design flow can also support Xilinx Virtex 4 and 5 FPGAs. Until the writing of this book in 2008, single-slice bus macros were proposed by Xilinx to support the connection between dynamic and static areas in Virtex 4 and 5 FPGAs. The single-slice bus macros can be placed anywhere within a PRR, that is, the restriction on placing a bus macro on the boundary of a PRR is no longer enforced.

As shown in Figure 7.4, a DPRS contains two PRRs (`PRR_a` and `PRR_b`), each of which can be configured into two different PRMs. The two PRRs are in the dynamic area, while the remaining design is in the static area. For communication between a PRR and the static area, it is necessary to place CLB bus macros on the boundary of the PRR. It is possible to configure the DPRS architecture into any one of the four combinations of PRMs, that is, {PRM_a1, PRM_b1}, {PRM_a1, PRM_b2}, {PRM_a2, PRM_b1}, and {PRM_a2, PRM_b2}, as illustrated in the directory structure given in Figure 7.5. Thus, it is strongly recommended that the EA PR design flow should follow this directory structure. The use of the folders is described in the following.

- The **hdl** directory is used to store all HDL files, including the top-level design, the static designs, and the PRMs.
- The **synth** directory is used to store all the synthesis project files. The **top** folder includes the global logic hardware components and the top-level HDL file, where the static designs and PRMs are instantiated as black boxes. All the synthesis projects for the static designs are located in the **static** directory while the PRMs are located in the corresponding PRR folders. For the DPRS architecture in Figure 7.4, the **prr_a** and **prr_b** directories individually contain two different PRMs.
- The **non_pr** directory includes all synthesized designs for the DPRS architecture. Though the DPRS architecture illustrated in Figure 7.4 can be configured into four different combinations of PRMs, it is strongly

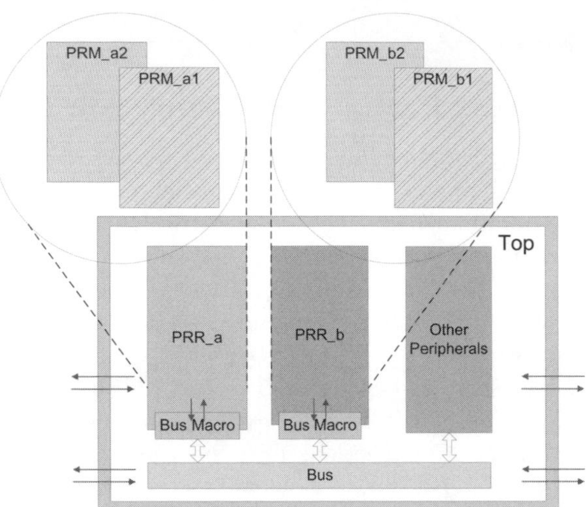

Figure 7.4: PR architecture.

recommended that a designer should generate the full bitstream of the whole architecture for each different PRM combination. The purpose is to validate the functional correctness of the full system as early as possible before going through the EA PR design flow.

- The **pr** directory includes the main PR implementation flow. All the synthesized designs, including the top-level design, the static designs, and each of the PRMs in the **synth** directory, must be copied to this **pr** folder. The **static** directory is used to implement the static design, while each **prm** directory is used to implement a PRM. The **merge** directory is used to collect all the implemented designs in the **static** and **prm_xx** directories, and then generate the static_full bitstream for the static designs and the partial bitstreams for all the PRMs.

As shown in Figure 7.6, the EA PR design flow can be divided into several main steps. The *Design Partitioning and Synthesis* step is used to modify the HDL description for fitting the DPRS architecture and to synthesize the designs into the `hdl` and `synth` directories. In the *Design Budgeting* step, the synthesized top-level design needs to meet the physical FPGA design constraints, including the location of bus macros and global logics. The *Non-PR Design* step is used to integrate the synthesized top-level design, the synthesized static designs, all combinations of the synthesized PRMs, and the design constraints provided by the *Design Budgeting* step for generating the full design bitstreams. By configuring each full bitstream, the functional correctness of DPRS can be validated before going through the EA PR design flow. When the validation of the DPRS architecture in the *Non-PR Design* step is finished,

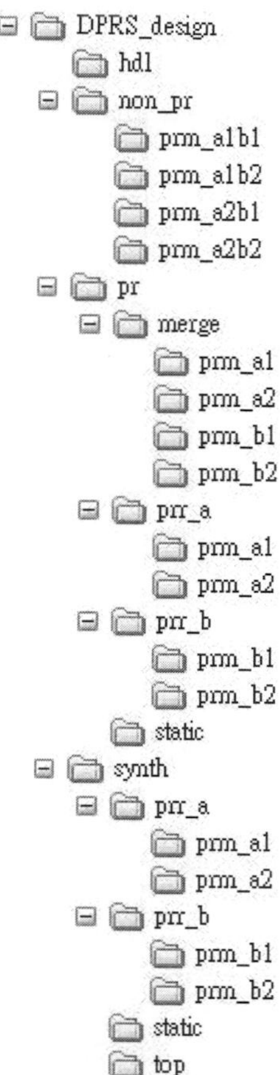

Figure 7.5: EA PR directory structure.

the final DPRS implementation needs to use the PR implementation design flow, including *Top-Level Implementation*, *Static Logic Implementation*, *PR Block Implementation*, and *Merge* steps.

Before venturing on the track of implementing of a DPRS architecture, the Xilinx ISE tool must be updated to the PR version. The PR implementation tool can be downloaded from the Xilinx website for the partial reconfiguration early access software tools [211]. After getting the PR implementation tool, the files must be copied to the installed ISE directory, then the following command must be issued on the command line to update your ISE tool.

```
xilperl PRinstall.pl PRfiles.txt
```

After the PR implementation tool is successfully installed, the `XInfo System`

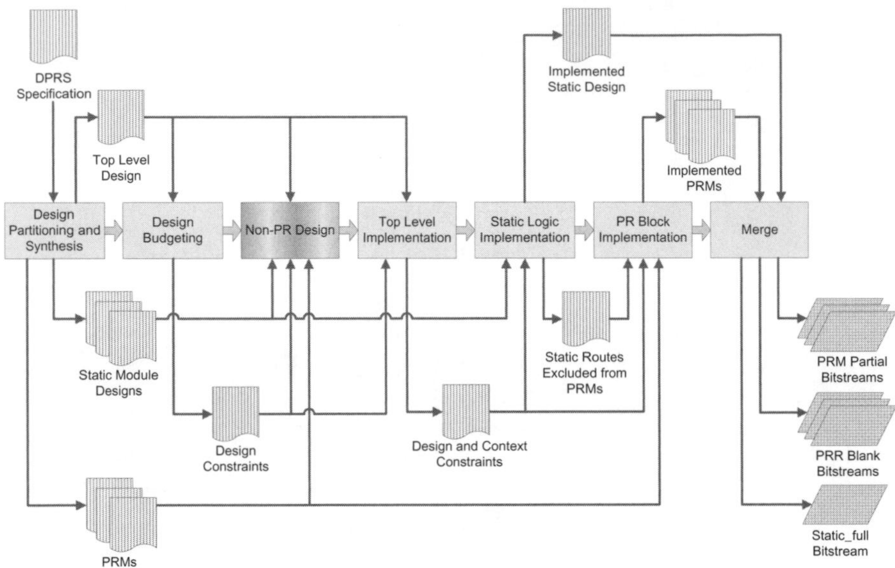

Figure 7.6: EA PR design flow.

`Checker` of the ISE tool can be used to check if the ISE version is updated to the PR version. In the following subsections, the main steps will be illustrated in detail.

7.2.1 Design Partitioning and Synthesis

This step is to define the dynamic area and the static area in a DPRS. First, the HDL descriptions need to be rewritten according to the DPRS architecture design, and then all hardware designs need to be synthesized by using the ISE tool. The DPRS design must be hierarchical, which is separated into the top-level design, the static design, and the PRMs. All I/O instantiations, clock primitive instantiations (BUFGs and DCMs), signal declarations, and bus macro instantiations must be combined in the top-level design, while the static designs and PRMs are instantiated as black boxes. All global logic resources are integrated with the HDL file for the top-level design to be synthesized for generating the top-level design netlist.

Compared to the top-level design, the static designs and the PRMs are its submodule designs, and thus the clock and reset related primitives cannot be included. Before the static designs and the PRMs are synthesized, the I/O buffers cannot be inserted because all the I/O signal transfers can only be through the top-level design. Therefore, the `Add I/O Buffers` item (*Synthesize - XST → Properties → Xilinx Specific Options → Add I/O Buffers*) needs to be disabled when static designs and the PRMs are synthesized using the

XST of the Xilinx ISE tool. Moreover, all PRMs in a PRR must be defined the same port definitions and the entity names because the communication with the static designs or other PRR is based on the ports of the PRR but not on that of the PRMs. This shows that every PRM in a PRR has the same interface for interacting with the static designs or other PRRs.

All signals of a PRR cannot directly interact with the external hardware designs. Here, a specific hardware component, namely the bus macro provided by Xilinx, must be used. The bus macros provided by Xilinx are the hard macro designs, and thus they need to be defined as a submodule for inserting in the top-level design. Before the top-level design, which is modified to be a DPRS architecture, is synthesized, the bus macro sources have to be copied to the same directory. There exist several different types of bus macros to fit the requirements for the design architecture, which include the direction, synchronicity, physical width and the FPGA families. The bus macro naming conversion can be described as *busmacro_device_direction_synchronity_width.nmc* for CLB bus macros and *busmacro_device_synchronity.nmc* for single-slice bus macros, as shown in the following table.

Table 7.1: CLB Bus Macros

Name	Type	Description
Device	xc2v	Virtex-II
	xc2vp	Virtex-II Pro
	xc4v	Virtex 4
	xc5v	Virtex 5
Direction	r2l	Right-to-left
	l2r	Left-to-right
	b2t	Bottom-to-top (Virtex 4 only)
	t2b	Top-to-bottom (Virtex 4 only)
Synchronicity	sync	Synchronous
	async	Asynchronous
Width	wide	Wide bus macro
	narrow	Narrow bus macro

The left side of Figure 7.7 shows a simple sample for illustrating the CLB bus macro usage, where four different directions of the CLB bus macros are used to pass the signals into the PRR. For the width of CLB bus macros as shown in the right side of Figure 7.7, the narrow bus macros are two CLBs wide while the wide bus macros are four CLBs wide. However, the width of a CLB bus macro is only 8-bit. In order to extend the bus width of CLB bus macros in a CLB row, three wide CLB bus macros can be nested in the same CLB row for achieving 24-bit bus bandwidth, as shown in Figure 7.8. Compared to the CLB bus macros, the single-slice bus macros are not

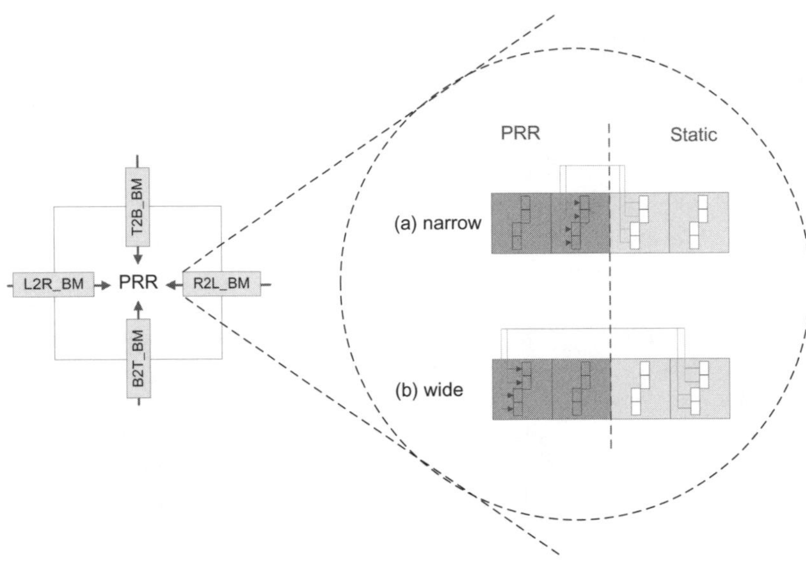

Figure 7.7: CLB bus macros.

directional and only support synchronicity type. Furthermore, the single-slice bus macros can be placed anywhere within the PRR instead of the placement in the boundary of a PRR for the CLB bus macros. As of the writing of this book in 2008, the single-slice bus macros can support the Virtex 4 and Virtex 5 FPGAs. In the following subsections, we illustrate every step of the PR implementation flow.

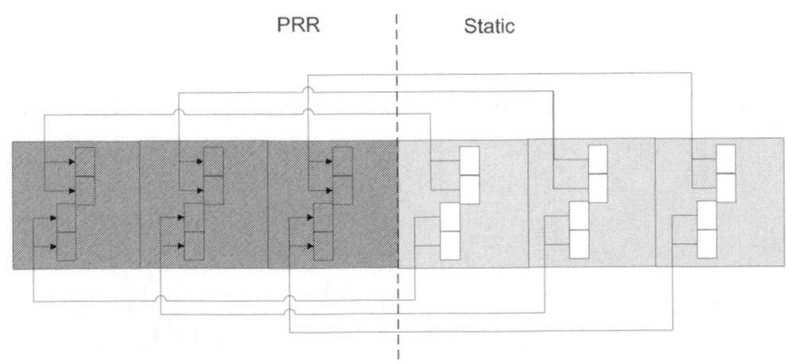

Figure 7.8: Three nested CLB bus macros.

7.2.2 Design Budgeting

After the top-level design, static designs, and PRMs are synthesized, the synthesized top-level design needs to be used for the budget of the physical logic components on the FPGA. Many design constraint items, including AREA GROUP, AREA RANGE, MODE and LOC, must be adopted to budget the DPRS architecture. Then, all the physical design constraints are defined in the *User Constraint File* (UCF) for the implementation of the DPRS architecture.

- **AREA GROUP** is used to group several instantiations in the static design or define a PRR on the FPGA. Here is a simple example to show the use of AREA GROUP.

```
INST "DPR_core_0" AREA_GROUP = "DPR_Module_0";
INST "lcd_optional" AREA_GROUP = "static";
INST "spi_eeprom" AREA_GROUP = "static";
INST "dsocm_bram" AREA_GROUP = "static";
```

- **AREA RANGE** is used to define the rectangular size of the PRR. The rectangle PRR is defined from the lower left corner (minX, minY) to the upper right corner (maxX, maxY). Moreover, the PRR must be carefully placed on the FPGA according to the physical placement constraints, as illustrated in Section 7.1. Here, we use an example to illustrating the use of the AREA RANGE.

```
AREA_GROUP "DPR_Module_0" RANGE=SLICE_X12Y96:SLICE_X31Y159;
AREA_GROUP "DPR_Module_0" RANGE=TBUF_X12Y96:TBUF_X30Y159;
AREA_GROUP "DPR_Module_0" RANGE=MULT18X18_X1Y12:MULT18X18_X2Y19;
AREA_GROUP "DPR_Module_0" RANGE=RAMB16_X1Y12:RAMB16_X2Y19;
```

- **MODE** is used to define the PRR property for preventing the errors during the static design and the PRM implementation. The **MODE** of every PRR must be defined as follows.

```
AREA_GROUP "DPR_Module_0" MODE=RECONFIG;
```

- **LOC** is used to set the I/O pins, the global clock primitives, and the bus macros. However, there are some constraints for the placement of CLB bus macros to follow. The CLB bus macros can only be placed on the boundary of a PRR, where the x-coordinate of the bus macro equals the maximum x-coordinate or is two less than the minimum x-coordinate of the PRR when the bus macros are in the right or left boundary, respectively. In contrast to the CLB bus macros, the single-slice bus macros can be placed anywhere within a PRR. Here is an example for the use of LOC.

```
NET "fpga_0_RS232_Uart_RX_pin" LOC = F14;
INST "sys_clk_s_BUFG" LOC = BUFGMUX6S;
INST "BM4DPR0_dataout" LOC = SLICE_X30Y122;
```

For easier budgeting of a DPRS design, Xilinx provides a graphic interface tool, called PlanAhead, for designers. Some PR implementation steps are integrated in the tool so that designers can use its graphic interface to implement the DPRS architecture. This section focuses on the illustration for the EA PR design flow, which can help you clearly understand every step of the EA PR design flow. The PlanAhead tool will be adopted to implement a simple DPRS design in Section 7.3, where the use of the PlanAhead tool will be illustrated in detail.

7.2.3 Non-PR Design

Before implementing the DPRS architecture, this step is strongly recommended. The final generated bitstreams of this step include the complete DPRS functionalities. All hardware designs, including the top-level design, static designs, and different combinations of PRMs, are integrated to generate full bitstreams without going through the PR implementation design flow. This step is used to validate the functional correctness of a DPRS before the final DPRS is created. The bitstream generation is similar to the common flow except the UCF file defined by the *Design Budgeting* step needs to be inserted.

Before the UCF file is used to generate the bitstream, the MODE = RECONFIG constraints for the PRRs must be commented or removed and the AREA GROUP constraints for each PRR need to be disabled, such as the following case.

```
#AREA_GROUP "DPR_Module_0" MODE=RECONFIG;
INST "DPR_core_0" AREA_GROUP = CLOSED;
```

The top-level design, the static designs, one combination of the PRMs, and the modified UCF file must be copied to the **non_pr** directory. Then the following commands are used to generate the bitstream.

>non_pr**ngdbuild** `-uc top.ucf top.ngc`

```
>non_pr\map top.ngd
>non_pr\par top.ncd top_nonpr_routed.ncd
>non_pr\bitgen -w top_nonpr_routed.ncd
```

For the DPRS architecture in Figure 7.4 and the directory structure in Figure 7.5, four different combinations of PRMs in the DPRS need to be individually generated by following the commands. After validating the functional correctness of every combination of PRMs, the PR implementation design flow can be used to generate the static_full bitstream for static area and the partial bitstreams for the PRMs.

7.2.4 Top-Level Implementation

This step is the first step of the PR implementation flow for constructing the DPRS architecture. The design constraint file (UCF file) and the synthesized top-level design are used to generate a top-level context file, including the placement for the I/O instantiations, the clock primitive instantiations, the bus macro instantiations, and the range of PRRs on the FPGA. The synthesized top-level design and the used bus macro files need to be copied to the **pr** directory, and then the following commands are used in the command line.

```
>pr\ngdbuild -uc top.ucf -modular initial top.ngc
```

7.2.5 Static Logic Implementation

The purpose of this step is to place and route the static designs on the used FPGA. The synthesized static design, the used bus macros, and the design constraint file (UCF file) are copied to the **static** directory. Referring to the top-level context file, the synthesized static design is implemented for the DPRS architecture using the following commands in order.

```
>pr\static\ngdbuild -uc top.ucf -modular initial ..\top.ngc
>pr\static\map top.ngd
>pr\static\par -w top.ncd top_static_routed.ncd
```

In contrast to the modular-based-design partial reconfiguration flow [95], the implementation of the static designs can use the logic resources of the PRRs for routing. After the static logic implementation step finishes, a specific file, namely "`static.used`," is generated to record which routing resources of the PRRs are used for the static design. Therefore, the PR block implementation step can avoid using the routing resources that are used by the static design; thus, the system execution will not be affected by partial reconfiguration.

7.2.6 PR Block Implementation

This step is used to separately implement each PRM for the DPRS architecture in its corresponding PRR. For the DPRS architecture in Figure 7.4, it is suggested that four PRMs are individually implemented in four corresponding directories as shown in Figure 7.5. Similar to the static logic implementation step, the design constraint file (UCF file), the bus macro files, and the synthesized PRM are copied to the directory. Moreover, the "`static.used`" file needs to be copied to the directory and renamed "`arch.exclude`." Referring to the "`arch.exclude`" file, the PRM implementation can use the remaining routing resources of the PRR without affecting the use of the static design. After preparing all the required files in the directory, the following commands are used in order in the command line.

>pr\prr_a\prm_a1**ngdbuild** `-uc top.ucf -modular module -active prm_a1 ..\..\top.ngc`
>pr\prr_a\prm_a1**map** `top.ngd`
>pr\prr_a\prm_a1**par** `-w top.ncd prm_a1_routed.ncd`

>pr\prr_a\prm_a2**ngdbuild** `-uc ..\top.ucf -modular module -active prm_a2 ..\..\top.ngc`
>pr\prr_a\prm_a2**map** `top.ngd`
>pr\prr_a\prm_a2**par** `-w top.ncd prm_a2_routed.ncd`

For implementing the PRMs corresponding to PRR_b, we can use similar commands as described above.

7.2.7 Merge

The final step of the PR implementation flow is to merge the top-level design, the static designs, and the PRMs. The implemented static design and the implemented PRMs are copied to the corresponding PRM directory and then the static_full bitstream and the partial bitstreams for each PRM can be generated using the following commands.

>pr\merge\prm_a1**pr_verifydesign** `top_static_routed.ncd prm_a1_routed.ncd`
>pr\merge\prm_a2**pr_assemble** `top_static_routed.ncd prm_b1_routed.ncd`

In the same way, the remaining partial bitstreams for the PRMs can be generated using similar commands. Additionally, the specific partial bitstreams, namely blank bitstreams, are also generated to clear the PRM in the PRR. The following are the final bitstreams after going through the EA PR design

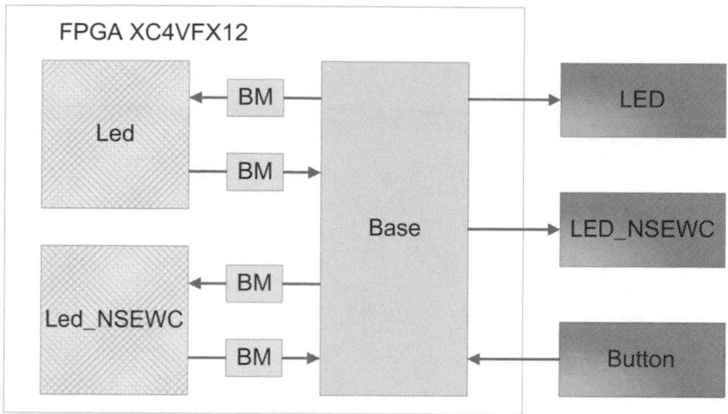

Figure 7.9: Simple PR LED control.

flow according to the DPRS architecture in Figure 7.4.

```
static_full.bit
prr_a_blank.bit (used to clear PRR_a)
prr_b_blank.bit (used to clear PRR_b)
prm_a1_routed_partial.bit (used to load PRM_a1 into PRR_a)
prm_a2_routed_partial.bit (used to load PRM_a2 into PRR_a)
prm_b1_routed_partial.bit (used to load PRM_b1 into PRR_b)
prm_b2_routed_partial.bit (used to load PRM_b2 into PRR_b)
```

After successfully generating the bitstreams, the iMPACT tool is used to download the static_full.bit bitstream and then each of the partial bitstreams for validating their functionalities.

7.3 Creating Partially Reconfigurable Hardware Design

To illustrate the detailed implementation steps in system design flow, we present a simple partially reconfigurable architecture, namely a LED controller. The DPRS architecture is implemented based on the EA PR design flow. The Xilinx ML403 FPGA development board [212] is used as the design platform and the design tools include the Xilinx ISE version 8.2.01i_PR_5 and PlanAhead version 8.2.4. PlanAhead allows a system design to be implemented more easily, instead of the original method that requires the use of many commands. The LED controller is illustrated in Figure 7.9.

In this LED controller design, there is a total of nine LEDs, namely four general-purpose LEDs and five location LEDs. The location LEDs are called

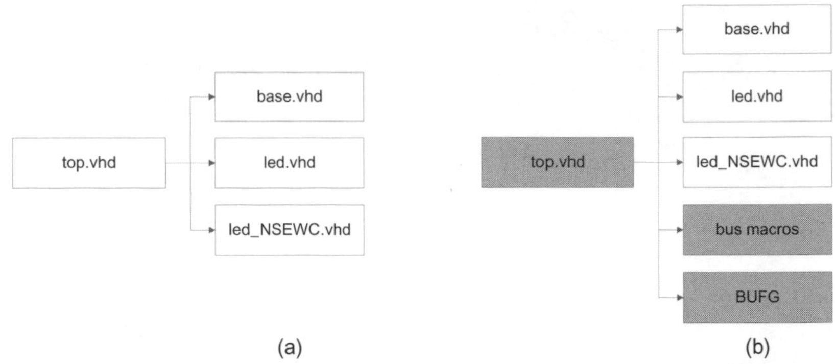

Figure 7.10: Simple PR LED HDL.

North, South, East, West, and Center LEDs, and they are triggered by five corresponding buttons. The LED controller design has a hierarchical DPRS architecture, consisting of a top-level design (`top.vhd`) and three submodule designs, namely, `base.vhd`, `led.vhd`, and `led_NSEWC.vhd`, as illustrated in Figure 7.10(a).

As illustrated in Figure 7.9, the `Base` design block receives the inputs from the five buttons, which are then passed on to both the `Led` and `Led_NSEWC` design blocks. The `Led` block maps the input data $\{N, S, E, W\}$ to $\{0, 1, 2, 3\}$ and the `Led_NSEWC` block maps $\{N, S, E, W, C\}$ to $\{N, S, E, W, C\}$. The exact I/O mappings are determined by the designer at design time for each block. The mapped data are then sent from these two blocks to the Base design, which then controls all nine LEDs. In this case study, the `Led` and `Led_NSEWC` design blocks are implemented as PRMs. There are two versions for each of the two blocks, differing only in the mapping functions. By reconfiguring different PRMs into the two PRRs, even though the input data from the button are the same, yet different LEDs will be switched on, based on the configured mapping functions (PRMs). Next, we will illustrate, step by step, how to modify the LED control design into a DPRS architecture.

In the EA PR design flow as described in Section 7.2, the first step is to modify the top-level HDL description as follows. A global buffer (BUFG) and bus macros have to be inserted in the top-level design, as shown in Figure 7.10(b). Some rules that can help designers to clearly understand the HDL-description modification are as follows. Figure 7.11 illustrates how these four rules are applied to the LED controller example.

- **Rule 1**: The PRMs cannot directly connect to the FPGA I/O pins in the top-level design. The signals of PRMs need to be passed to/from the static designs or connected through the bus macros for connecting to the FPGA I/O pins.
- **Rule 2**: Every PRM must include a signal for triggering (clock).

- **Rule 3**: The clock signal cannot be passed directly to a PRM. The global buffer must be used to pass the clock signal to the PRMs.

- **Rule 4**: A PRM must use the bus macros to interact with the static design or another PRM.

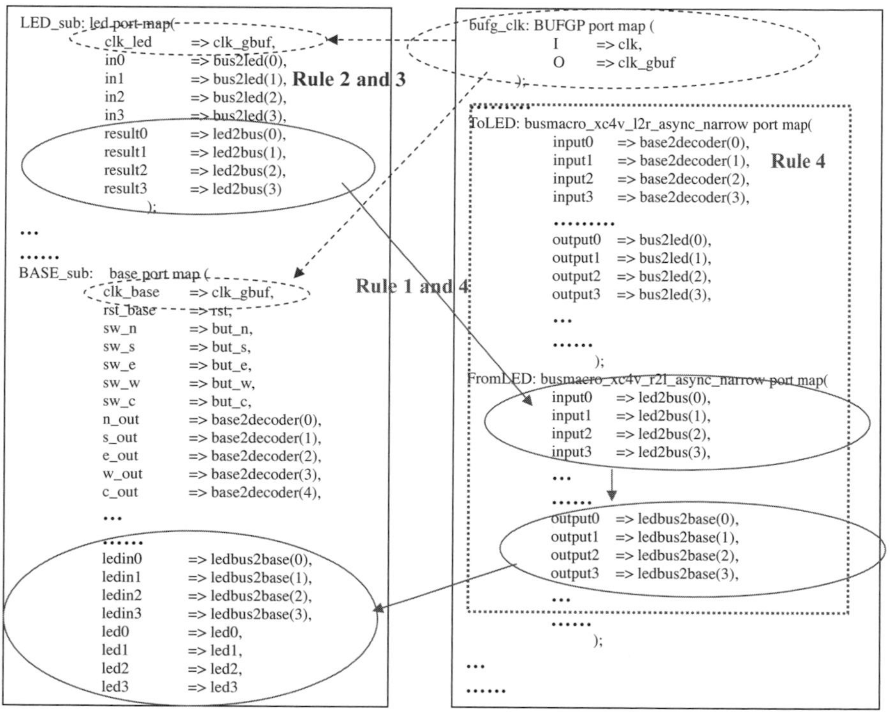

Figure 7.11: Modified top-level HDL.

After modifying the top-level HDL description, the four HDL files shown in Figure 7.10(a) must be separately synthesized by following the approach described in Section 7.2.1. The top-level design includes the bus macros and the I/O buffer. Moreover, the top-level design for the FPGA pin assignments needs to be defined in the UCF file, as shown in Figure 7.12. However, before the PR implementation flow is used, the synthesized top-level design, the bus macros, and the UCF file need to be copied to a directory, without including the synthesized `Base`, `LED`, and `LED_NSEWC` designs. In the following, the use of the PlanAhead tool is illustrated for implementing the DPRS architecture.

Top-Level HDL file

```
entity top is
    port(
        clk     : in std_logic;
        rst     : in std_logic;
        but_n   : in std_logic;
        but_s   : in std_logic;
        but_e   : in std_logic;
        but_w   : in std_logic;
        but_c   : in std_logic;
        led0    : out std_logic;
        led1    : out std_logic;
        led2    : out std_logic;
        led3    : out std_logic;
        ledN    : out std_logic;
        ledS    : out std_logic;
        ledE    : out std_logic;
        ledW    : out std_logic;
        ledC    : out std_logic
    );
end top;
```

UCF file

```
# Switch buttons
NET "but_w" LOC = E9;
NET "but_s" LOC = A6;
NET "but_n" LOC = E7;
NET "but_e" LOC = F10;
NET "but_c" LOC = B6;

#Led lights
NET "led3" LOC = A12;
NET "led2" LOC = A11;
NET "led1" LOC = G6;
NET "led0" LOC = G5;

#NSEWC led lights
NET "ledN" LOC = E2;
NET "ledE" LOC = E10;
NET "ledS" LOC = A5;
NET "ledW" LOC = F9;
NET "ledC" LOC = C6;

#Reset and clock
NET "rst" LOC = D6;
NET "clk" LOC = AE14;
```

Figure 7.12: UCF file for LED control design.

◇ Creating a PlanAhead project

1. Start PlanAhead and create a new project.
2. Name and locate the project.
3. Import the synthesized top-level design (`top.ngc`).
4. Select the FPGA type, which is XC4VFX12-FF668-10.
5. Import the UCF file, which includes the I/O pin assignments, and finally click "Finish."

◇ Enabling PlanAhead-PR

This step is required because the PlanAhead tool does not automatically enable the partial reconfiguration feature. Enter the following command into the PlanAhead Tcl console for enabling the partial reconfiguration features.

```
hdi::param set -name project.enablePR -bvalue yes
```

◇ Floorplan

After the synthesized top-level design is inserted into the PlanAhead project, the *Netlist* view should be as shown in Figure 7.13.

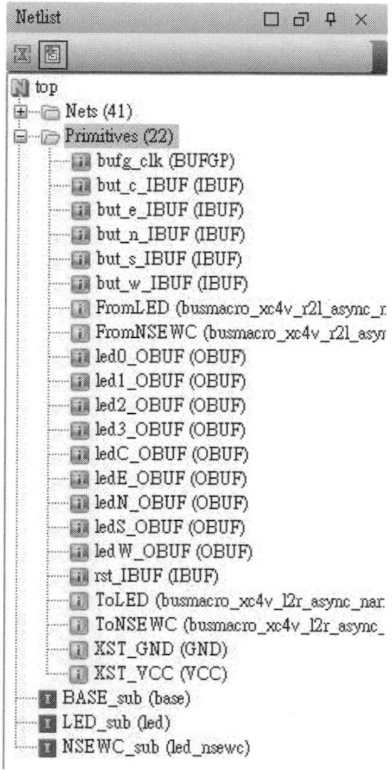

Figure 7.13: Netlist windows of PlanAhead.

1. Select and right-click the `Base` design in the *Netlist* view, and select "New Pblock" and use the default name `pblock_base_sub` for the following constraint in the UCF file.

   `INST "BASE_sub" AREA_GROUP = "pblock_BASE_sub";`

2. Select and right-click the `LED` and `LED_NSEWC` designs in the *Netlist* view, then select "Draw Pblock" to draw a slice range in the *Device* view for the following constraints in the UCF file.

   `AREA_GROUP "pblock_NSEWC_sub" RANGE=SLICE_X30Y106:SLICE_X41Y127;`

   `AREA_GROUP "pblock_NSEWC_sub" RANGE=RAMB16_X2Y14:RAMB16_X2Y15;`

   `AREA_GROUP "pblock_NSEWC_sub" RANGE=DSP48_X0Y28:DSP48_X0Y31;`

```
AREA_GROUP "pblock_NSEWC_sub" RANGE=FIFO16_X2Y14:FIFO16_X2Y15;
INST "NSEWC_sub" AREA_GROUP = "pblock_NSEWC_sub";
AREA_GROUP "pblock_LED_sub" RANGE=SLICE_X4Y92:SLICE_X19Y127;
AREA_GROUP "pblock_LED_sub" RANGE=RAMB16_X0Y12:RAMB16_X1Y15;
AREA_GROUP "pblock_LED_sub" RANGE=FIFO16_X0Y12:FIFO16_X1Y15;
INST "LED_sub" AREA_GROUP = "pblock_LED_sub";
```

3. Mark the Pblocks `pblock_LED_sub` and `pblock_NSEWC_sub` as reconfigurable. In the *Physical Hierarchy* view, separately select the Pblocks `pblock_LED_sub` and `pblock_NSEWC_sub` and click the "Attributes" tab in the *Pblock Properties* view. Then, click the "Define New Attribute" button and select "MODE," set its value to "RECONFIG" for creating the following constraints in the UCF file.

```
AREA_GROUP "pblock_NSEWC_sub" MODE=RECONFIG;
AREA_GROUP "pblock_LED_sub" MODE=RECONFIG;
```

4. Select the location for each of the bus macros and the global buffer. Change the selection mode to "Create Site Constraint Mode" as shown in Figure 7.14.

Figure 7.14: Create Site Constraint Mode.

Drag the four bus macros and the global buffer to the following locations in the UCF file.

```
INST "ToLED" LOC = SLICE_X2Y114;
INST "FromLED" LOC = SLICE_X2Y120;
INST "ToNSEWC" LOC = SLICE_X28Y110;
INST "FromNSEWC" LOC = SLICE_X28Y120;
INST "bufg_clk" LOC = BUFGCTRL_X0Y3;
NET "clk" LOC = AE14;
```

◇ Design Rule Check

This step should be performed before implementation, because it can help a designer find errors in the HDL code or during floorplanning. Select the menu "Tools → Run DRC." Select "Specify rules to check" and turn on "Partial Reconfig."

◇ Netlist Export

After a design passes the DRC checker, it can be used for the PR implementation. Select the menu "File → Export Floorplan." Set the export mode to "Partial Reconfig." The synthesized top-level design, the new UCF file, and the bus macro files will be automatically copied to the top-level directory.

◇ Run Partial Reconfiguration Design Flow

The purpose of this step is to use the exported design to generate a static_full bitstream for the static design and a partial bitstream for each of the PRMs. The tool will automatically run ngdbuild, map, par, trace, and the PR assemble scripts to create the bitstreams. After the floorplan is exported, the PlanAhead project created should include the directories as shown in Figure 7.15. The synthesized static design, one of the `LED` PRMs, and one of the `LED_NSEWC` PRMs must be copied to the *static*, *pblock_LED_sub_CV*, and *pblock_NSEWC_sub_CV* directories, respectively.

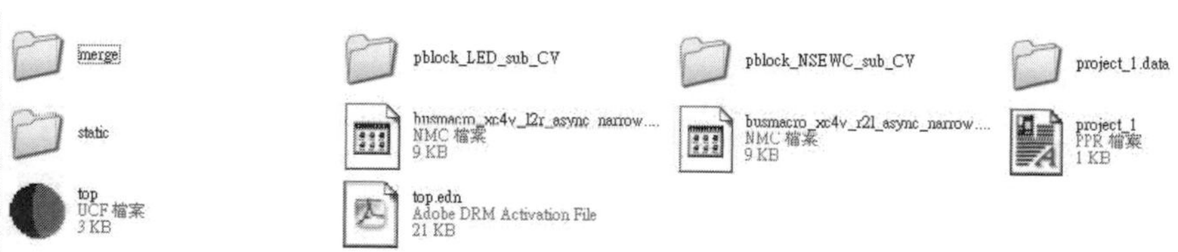

Figure 7.15: PlanAhead directory.

After the required sources are copied to the corresponding directories, start the flow wizard from the menu "Tool → Run Partial Reconfig." This will either create a script to be run from a command line or run in an interactive mode. Select the "Run Place & Route." Then run the implementation steps in order.

1. **Budgeting**
   In the budgeting phase, the UCF file is used to ensure the correctness of the design. Select the "Budgeting" item to start the partial reconfiguration flow. Accept the default setting for the step and click "Finish." The "Place & Route Output" window will appear. It should report 0 error to pass this phase.

2. **Static Logic Implementation**
   In this step, the static portion of the design is implemented. Select the "Static Logic Implementation" item and then accept the default setting for the step and click "Finish." The "Place & Route Output" window will appear. It should report 0 error to pass this phase.

3. **PR Block Implementation**
   In this step, each of the PRMs in the design is implemented. Select the "PR Block Implementation" item and then click the "Add" button to add the *pblock_LED_sub* and *pblock_NSEWC_sub* Pblocks. Accept the default setting for the step and click "Finish." The "Place & Route Output" window will appear. It should report 0 error to pass this phase.

4. **Assemble**
   In this step, the static designs and PRMs are merged to generate the corresponding bitstreams for configuring the FPGA. Select the "Assemble" item and then accept the default setting for the step and click "Finish." Finally, the bitstreams are generated in the *merge* directory.

In the same way, copy the other two PRM netlists to the *pblock_LED_sub_CV* and *pblock_NSEWC_sub_CV* directories. Then, go through the **PR Block Implementation** and the **Assemble** steps again to generate the partial bitstreams. The final generated bitstreams should include:

```
static_full.bit
pblock_led_sub_blank.bit (used to clear pblock_LED_sub)
pblock_nsewc_sub_blank.bit (used to clear pblock_NSEWC_sub)
pblock_led_sub_cv_routed_partial_A.bit
pblock_led_sub_cv_routed_partial_b.bit
pblock_nsewc_sub_cv_routed_partial_A.bit
pblock_nsewc_sub_cv_routed_partial_B.bit
```

Next, start the iMPACT tool to configure the bitstreams to the FPGA in order to check the correctness of the DPRS. First, the static_full.bit is configured into the FPGA. Each partial bitstream of *pblock_LED_sub* and *pblock_NSEWC_sub* pblocks are then configured into the FPGA. Click the FPGA push buttons to check which LEDs will be switched on. Next, reconfigure other partial bitstreams of *pblock_LED_sub* and *pblock_NSEWC_sub* pblocks into the FPGA to ensure the correctness of partial reconfiguration. After the partial reconfiguration is done, the LED results should be different.

This is a very simple DPRS architecture, which introduces a basic PR implementation method with a pure hardware design. In the next section, we will provide a hardware/software design for the DPRS architecture for more concrete understanding of the PR implementation.

7.4 Software-Controlled Partially Reconfigurable Design

After introducing the hardware only case study in Section 7.3, we will now explain how a hardware/software design can be enhanced into a DPRS architecture in this section. In this case study, not only the Xilinx ISE, and PlanAhead tools are used, but the Xilinx EDK tool is also used to create a hardware/software embedded system on a Xilinx FPGA device, where the EDK, ISE, and PlanAhead versions are 8.2.02i, 8.2.01i_PR_5, and 8.2.4, respectively.

In this DPRS design, the CRC and RSA designs will be implemented as PRMs and (re)configured into the same PRR. Different from the external configuration control employed in Section 7.3, the *internal configuration access port* (ICAP) will be used to (re)configure the CRC and RSA PRMs into the FPGA. Here, the specific APIs for the ICAP can be included in a software application to control the partial reconfiguration. In this way, an application can use the same hardware resources to perform different hardware functions according to the current configured PRM on the FPGA.

Recall that a PRR has a fixed I/O interface so all PRMs to be configured into the PRR also need to have the same interface. The original CRC and RSA designs had different I/O interfaces, so a common interface needs to be defined to allow them to share the same PRR. As a result, a software application can interact with the CRC or RSA designs through the common interface. As a solution, a partially reconfigurable template [93] is proposed for integrating the CRC and RSA design interfaces. The template consists of eight 32-bit input data signals, one 32-bit input control signal, four 32-bit output data signals, and one 32-bit output control signal. Moreover, the IPIF provided by Xilinx EDK is used for communication between the hardware and software designs.

The design shown in Figure 7.16 is a PowerPC-based architecture. The communication infrastructure is the IBM CoreConnect architecture, which includes a high-speed *Processor Local Bus* (PLB) and a low-speed *On-chip Peripheral Bus* (OPB). The RS232 controller is responsible for displaying the messages in the terminal, while the JTAG controller is responsible for externally configuring the FPGA. The partial bitstreams are stored in the CF card, while the ICAP is used to internally (re)configure the partial bitstreams into the FPGA. The clock signals are managed through the DCM component,

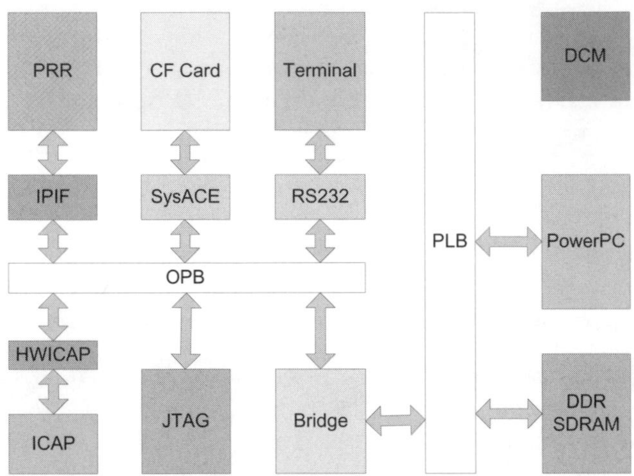

Figure 7.16: Software-controlled DPRS architecture.

and the IPIF is used to build a software-accessible interface for interacting with the PRR. All the hardware designs were implemented on the Xilinx ML310 development board [214].

To design a software-controlled dynamically partially reconfigurable system with ICAP, we recommend the basic design flow as illustrated in Figure 7.17. First, the base system design is built, and then a new hardware design is added to the top-level design, along with a communication interface, connected to the OPB bus, for receiving the data from the software application. After the generated full bitstream of the system is validated, the system design is then changed to a DPRS architecture, where the new hardware design is designed as a PRM now. The EA PR design flow is adopted to generate the static_full and partial bitstreams. By configuring these bitstreams into an FPGA their functional correctness can be validated. Finally, the ICAP is integrated into the system design such that partial reconfiguration can be performed without external control. The design flow includes several steps for validating the functional correctness such that design errors can be detected as early as possible, before the final system is created. In the following, we will illustrate step by step the implementation of the software-controlled partially reconfigurable design.

◇ Build Basic System Design

1. Create a new EDK project, and then select the Virtex-II Pro ML310 Evaluation Platform.

2. Set both data and instruction on-chip memory to 8 KB.

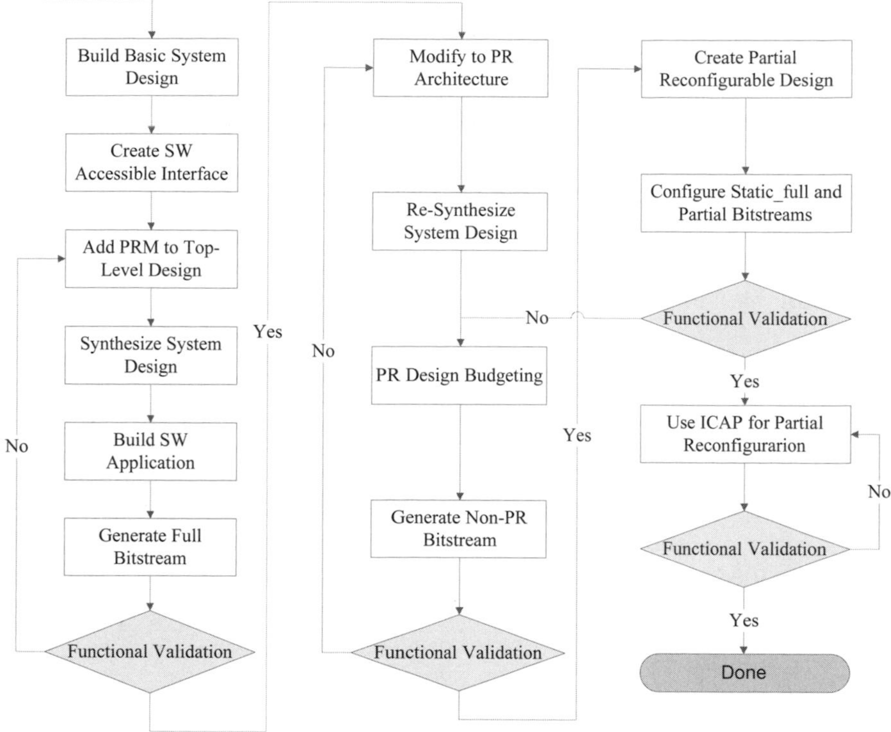

Figure 7.17: Software-controlled DPRS design flow.

3. Select `RS232_Uart`, `DDR_SDRAM_32M×64`, `SPI_EEPROM`, `LED_8Bit`, `LCD_OPTIONAL` and `SysACE_CompactFlash` peripherals, where `RS232_Uart`, `DDR_SDRAM_32M×64`, `SPI_EEPROM` and `SysACE_CompactFlash` peripherals must contain the interrupt support.

4. Set the memory size of PLB BRAM IF CNTLR to 128 KB.

5. In the IP catalog, expand the item "FPGA Reconfiguration" and then select `opb_hwicap` to attach to the OPB.

6. Use the "Generate Addresses" button to automatically assign the memory range for each hardware device.

◇ Create a Software-Accessible Interface

The purpose of this step is to use the IPIF to design a software-accessible hardware communication interface, connected to the PRR.

1. Select the menu item "Hardware" → "Create or Import Peripheral."
2. Select the item "Create template for a new peripheral," then attach the peripheral to the OPB. In the **pcores** directory, there will be two HDL files, including a wrapper for connecting the OPB and a user_logic design for integrating with the new peripheral. Here, sixteen software accessible registers are defined to connect to the partially reconfigurable template [93].
3. In the design, the slave registers 2 to 9 are connected to the input data ports, the slave registers 10 to 13 are connected to the output data ports, the slave register 14 is connected to the input control port, and the slave register 15 is connected to the output control port. The fourteen ports are added in the "user_logic.vhd" file to interact with the PRR as follows.

```
entity user_logic is
  generic
    ...
  port
  (
    -- ADD USER PORTS BELOW THIS LINE ------------------
    -- USER ports added here
    datain0_2prr :  out std_logic_vector(31 downto 0);
    ...
    datain7_2prr :  out std_logic_vector(31 downto 0);
    cntrl_in_2prr:  out std_logic_vector(31 downto 0);
    dataout0_prr:  in std_logic_vector(31 downto 0);
    ...
    dataout3_prr:  in std_logic_vector(31 downto 0);
    cntrl_out_prr:  in std_logic_vector(31 downto 0);
    -- ADD USER PORTS ABOVE THIS LINE ------------------
    -- DO NOT EDIT BELOW THIS LINE ---------------------
    ...
```

4. Select the menu item "Hardware" → "Import existing peripheral," and then give the peripheral a new version.

5. In the IP catalog, expand the item "Project Repository" and then select the new peripheral, which was created using the IPIF and which was attached to OPB. Finally, use the "Generate Addresses" button to automatically assign the memory range for each hardware device again.

6. Select the menu "Hardware" → "Generate Netlists" to generate the netlists. All HDL files of the peripherals will be located in the **hdl** directory.

◇ Add PRM to the Top-Level Design

In this step, the PRM needs to be inserted into the top-level design and connected to the software-accessible communication interface.

1. The RSA and the CRC designs are first integrated with the partially reconfigurable template [93] as shown in Figure 7.18. Separately synthesize the RSA and CRC designs using the ISE tool, and then copy one of the CRC and RSA designs to the **implementation** directory.

2. The HDL of the top-level design needs to be re-modified for connecting the communication interface and one of the RSA and CRC designs. Here are parts of the modified top-level HDL description.

```
comm_0 :  comm_0_wrapper
  port map (
    datain0_2prr => ipif2prr0_datain0,
    ...
    cntrl_in_2prr => ipif2prr0_cntrl_in,
    dataout0_prr2 => prr02ipif_dataout0,
    ...
    cntrl_out_prr => pr02ipif_cntrl_out,
    OPB_Clk => sys_clk_s(0),
    ...
  );
PRR_0:  PR_template0 port map(
    clk => sys_clk_s(0),
    reset => opb_OPB_Rst,
    datain0 => ipif2prr0_datain0,
    ...
```

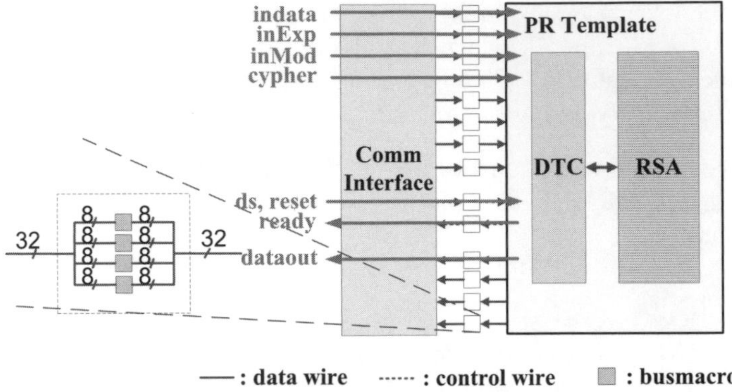

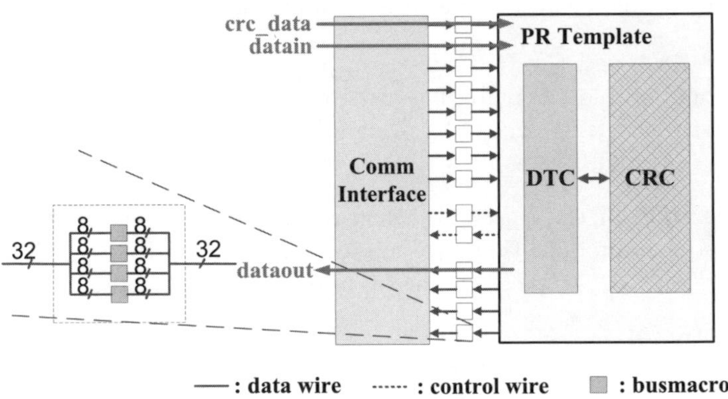

Figure 7.18: Integrate with the partially reconfigurable template.

```
    dataout0=> prr02ipif_dataout0,
    ...
    cntrl_in=> ipif2prr0_cntrl_in,
    cntrl_out=> prr02ipif_cntrl_out
  );
```

3. The top-level design needs to be re-synthesized due to the HDL modification. However, the menu item "Hardware" → "Generate Netlists" cannot be used again because the action will recover the modified HDL description. Here, the command line method is used. Change to the **synthesis** directory and then use "xst" to re-synthesize the top-level design as follows.

```
>edk_project\synthesis\xst -ifn system_xst.scr
```

4. After the communication interface is created using the IPIF, the **drivers** directory includes the standalone drivers for users to control the communication interface. The software application must include the header file of the communication interface to interact with the communication interface. Then, right-click software project and set "Build Project" for generating the software executable file (elf). Here are parts of the software application for the communication interface.

```
...
#include "comm.h"
...
int rsa(){
   ...
   COMM_mWriteSlaveReg2(XPAR_COMM_0_BASEADDR, data);
   COMM_mWriteSlaveReg3(XPAR_COMM_0_BASEADDR, exp);
   COMM_mWriteSlaveReg4(XPAR_COMM_0_BASEADDR, mod);
   xil_printf("Start RSA Encrypting!\n\r");
   COMM_mWriteSlaveReg14(XPAR_COMM_0_BASEADDR, 1);
   COMM_mWriteSlaveReg14(XPAR_COMM_0_BASEADDR, 0);
   while ( COMM_mReadSlaveReg15(XPAR_COMM_0_BASEADDR) != 1) ;
   xil_printf("Encryption Done!\n\r");
   cypher_text = COMM_mReadSlaveReg10(XPAR_COMM_0_BASEADDR);
   xil_printf("Cypher Text:  0x%08x \n\r", cypher_text);
}
```

5. When the top-level design is re-synthesized, click the menu item "Device Configuration" → "Download Bitstream." The software executable file and the full bitstream will be combined into a bitsream "download.bit." Then the iMPACT tool will be invoked to configure the bitstream.

◇ Modify to PR Architecture

When the full bitstream is validated, the top-level HDL description needs to be re-modified again. As illustrated in Section 7.2.1, the bus macros are added to act as a bridge between the communication interface and the PRR. Moreover, the clock signal cannot directly pass to the PRR, thus a global buffer is used to transfer the clock signal from the DCM to PRR. In this design, the bus macros used include `busmacro_xc2vp_l2r_async_narrow` and busmacro_xc2vp_r2l_async_narrow.

```
dcm_0 :  dcm_0_wrapper
  port map (
    CLKIN => dcm_clk_s,
    CLKFB => dcm_clk0_buf,
    CLK0 => dcm_clk0_buf,
    CLK90 => clk_90_s(0),
    CLK180 => open,
    ...
    LOCKED => dcm_0_lock,
    PSDONE => open,
    RST => net_gnd0
  );
bufg_0:  bufg
  port map (
    I => dcm_clk0_buf,
    O => sys_clk_s(0));
```

The top-level design needs to be re-synthesized again because the HDL description of the top-level design is modified. Moreover, the bus macro files (nmc format) are copied to the **synthesis** directory for synthesizing the top-level design. Change to the **synthesis** directory and use the following commands to re-synthesize the top-level design:
`>edk_project\synthesis\xst -ifn system_xst.scr`

◇ PR Design Budgeting

This step uses the top-level netlist for design budgeting. A PRR needs to be allocated, and bus macros and the global buffer need to be assigned to specific locations in the FPGA.

1. Create a **pr** directory to store all the required files. Place copies of the top-level netlist (system.ngc) and the RAM Memory Map (system.bmm and system_bd.bmm) files in the **implementation** directory, the base design constraint file (system.ucf) in the **data** directory, along with the bus macro files.

2. Open a new PlanAhead project and import the top-level netlist (system.ngc) and the base design constraint file (system.ucf). Create a pblock with "MODE = RECONFIG" for the RSA and CRC PRMs, and use a pblock to group the remaining hardware components. Lastly, the global buffer and the bus macros are assigned one by one to specific locations in the FPGA.

3. After passing the design rule check, export the the PR design floorplan for PR implementation.

4. Generate the full bitstream of the non-PR design and use the iMPACT tool to download the bitstream into the FPGA. However, the bitstream includes only the hardware design so the software design needs to be downloaded separately. Here, the XMD debug tool is used to download the software executable file by selecting menu "Debug" → "Launch XMD." Next, use the default setting and change to the software project directory. Finally, use the following commands to download the software executable file to the microprocessor.

   XMD% **rst** (Reset the processor)

   XMD% **dow `executable.elf`** (Download the executable file)

   XMD% **run** (Run the executable file)

◇ Create the Partially Reconfigurable Design

After the floorplan is exported, the PlanAhead project should include three main directories, including **static**, **pblock_PRR_CV**, and **merge**. Then, all netlists, except for the PRM and top-level netlists, are copied to the **static** directory, while one of the PRM netlists (PR_template0.ngc) is copied to the **pblock_PRR_CV** directory. Additionally, the RAM Memory Map (system.bmm and system_bd.bmm) files are copied to the **merge** directory for the bitstream generation.

When all required files are copied to the corresponding directories, enable the PlanAhead-PR to run the EA PR design flow for generating the static_full and partial bitstreams. For generating the partial bitstream for another PRM, copy that PRM netlist to the **pblock_PRR_CV** directory and then run the *PR Block Implementation* and *Merge* steps again. The final bitstreams should include:

```
static_full.bit
pblock_prr_blank.bit
pblock_prr_cv_routed_partial_RSA.bit
pblock_prr_cv_routed_partial_CRC.bit
```

Use the iMPACT tool to download the static_full bitstream and the partial bitstream of RSA. Similarly, use XMD to download the software executable for validation. After reconfiguring the partial bitstream of CRC, run the software executable file again to check if the CRC design can work correctly.

◇ Use ICAP for Partial Reconfiguration

When the partial reconfiguration through the JTAG can correctly work, in this step we use the ICAP to partially reconfigure the FPGA. By using the driver of ICAP, the software application can dynamically reconfigure the FPGA without needing any external support.

1. Copy partial bitstreams to the CF card.

2. Select the menu item "Software" → "Software Platform Settings," and then select the **xilfatfs** library for accessing the data in the CF card. Select the menu item "Software" → "Clean Libraries" and then select "Generate Libraries and BSPs" for re-generating the software libraries.

3. The header files for data access for the CF card and the ICAP driver are included first in the software application. Then, save the partial bitstream in the CF card and use ICAP to reconfigure the FPGA. Here are parts of the software application source for reference.

```
#include "sysace_stdio.h"
#include "xhwicap_i.h"
...
char RSA[] = "rsa.bit";
char CRC[] = "crc32.bit";
XHwIcap InstancePtr;
...
int configRSA(){
  void *fd;
  fd = sysace_fopen(RSA, "r"); // Open RSA partial bitstream
  ...
  if( reprogram(fd, &InstancePtr) == 0 )
  ...
  sysace_fclose(fd);
}
...
int reprogram (void *fd, XHwIcap *Instance){
  ...
  XHwIcap_CommandDesync(Instance);
  while (1) {
```

```
        bytes = sysace_fread(buf, 1, BLOCKSIZE_BYTES, fd);
        //BLOCKSIZE_BYTES is the partial bitstream size
        //Use a buffer to store the partial bitstream of the CF card
        if (!bytes)
          break;
        }else {
          XHwIcap_SetConfiguration(Instance, buf, BLOCKSIZE_WORDS);
          //Use ICAP for partial reconfiguration
          //Every partial reconfiguration is one word (32-bit)
          if (bytes != BLOCKSIZE_BYTES)
            break;
    ...
  }
```

Different from the partial reconfiguration described in Section 7.3, this case study for partial reconfiguration includes both hardware and software designs. Furthermore, the software application can use the ICAP driver to reconfigure the partial bitstreams into the FPGA at run-time. Take the design for example. While the system is running, there is a requirement for the hardware RSA design, the currently unused CRC PRM in the PRR can be partially reconfigured into be the RSA PRM without switching off the system. Compared to traditional embedded systems, the dynamically adaptive capability can significantly enhance the system performance and flexibility. Within the limited logic resources, the system can include multiple hardware functions, whose total amount of required logic resources is much more than that available in the FPGA device.

7.5 Operating System for Reconfigurable Systems

This section describes the last case study for the DPRS architecture. Section 7.3 introduced a purely hardware partially reconfigurable system, while Section 7.4 described a software-controlled dynamically partially reconfigurable system. In this section, we will use the same DPRS architecture described in Section 7.4. However, the DPRS architecture will be integrated with an embedded operating system in place of the standalone software application, where such an operating system can be called *Operating System for Reconfigurable Systems* (OS4RS).

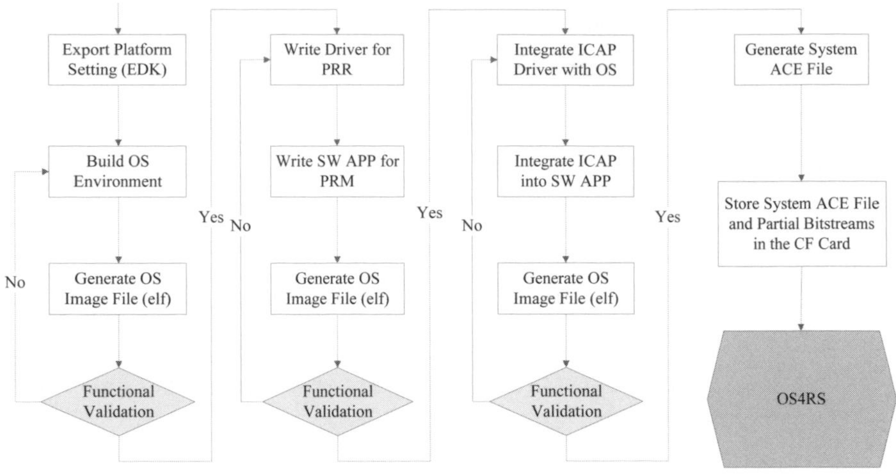

Figure 7.19: OS4RS design flow.

The OS4RS includes not only the basic OS management but also the PRMs. The PRMs in the OS4RS is akin to the software task in a conventional OS, and thus they can be called *hardware tasks*. However, the CRC and RSA designs are both integrated with the same partially reconfigurable template [93] such that a driver is designed for a PRR and not for the PRMs. The driver only provides an interface for the software application to interact with the PRR, where the functionality of software accessible registers will be different according to the configured PRM. As shown in Figure 7.18, the software accessible register 2 is mapped to the first input data ports of both the RSA and CRC designs. The driver is only responsible for writing the data to the software accessible register 2. However, the data definition for the first input data port should be different according to the definition of the configured PRM. The software application must be aware of this interactive method such that it can control the PRM through the same driver.

The OS4RS design flow is illustrated in Figure 7.19. The establishment of the DPRS architecture is similar to that in Section 7.4. The EDK project needs to export the information on the system environment, such as the memory size and memory mapping for each device. For each PRR, when using IPIF to design a communication interface, a *physical driver* for the standalone design is generated by the Xilinx EDK tool. On top of this physical driver, we need a *virtual driver* for each PRM. All the physical PRR drivers and the virtual PRM drivers are configured into the OS4RS kernel. Then, the iMPACT tool is used to download the non-PR full bitstream, as described in Section 7.4. Finally, we use the XMD tool to download the OS image file to the board for validation. After the functional correctness is validated, the ICAP driver proposed by John Williams [209] is integrated into the OS4RS kernel. The partial bitstreams and the software executable files are stored in the **root** directory of the file system, and then the ICAP driver is used for partial reconfiguration. Here, uClinux is used to build the OS4RS. The detailed steps of the OS4RS design flow are illustrated in the following.

◇ Export Platform Setting

Compared to the standalone design, this step rebuilds the software libraries for OS integration.

1. Select the menu item "Software" → "Software Platform Setting." In the *Processor Settings* window, select the cpu_ppc405 as the processor. In the *OS & Library Setting* window, select the "linux_mvl31" for the OS item.
2. Select the menu item "Software" → "Clean Libraries," and then select "Generate Libraries and BSPs" for re-generating the software libraries.
3. After the library generation finishes, two directories, including the **arch** and **driver**, are created and located in the directory "*edk_project\ppc405_0\libsrc\linux_mvl31_v1_00_a\linux.*"

◇ Build OS Environment

The OS4RS is based on the μClinux OS, whose kernel version is 2.4.26. The cross-compiler for the PowerPC processor needs to be installed in the Linux development environment. In the following, the integration between the hardware architecture created by the EDK tool and the μClinux OS will be discussed in detail.

1. Use the **arch** and **driver** directories that were generated in the previous step, to replace the directories in the kernel source with the same names.
2. In the main directory of the kernel source, use the command "**make menuconfig**" for kernel setting. For the item "General Setup," select "Default bootloader kernel arguments" and set "Initial kernel command string" to "`console=ttyS0,9600 root=/dev/xsysace/disc0/part3 rw`" for locating the file system in the third partition of the flash card as shown in Figure 7.20.
3. Because the target microprocessor is PowerPC, the main Makefile must be modified as follows.

```
ARCH := ppc
CROSS_COMPILE = powerpc-405-linux-gnu-
```

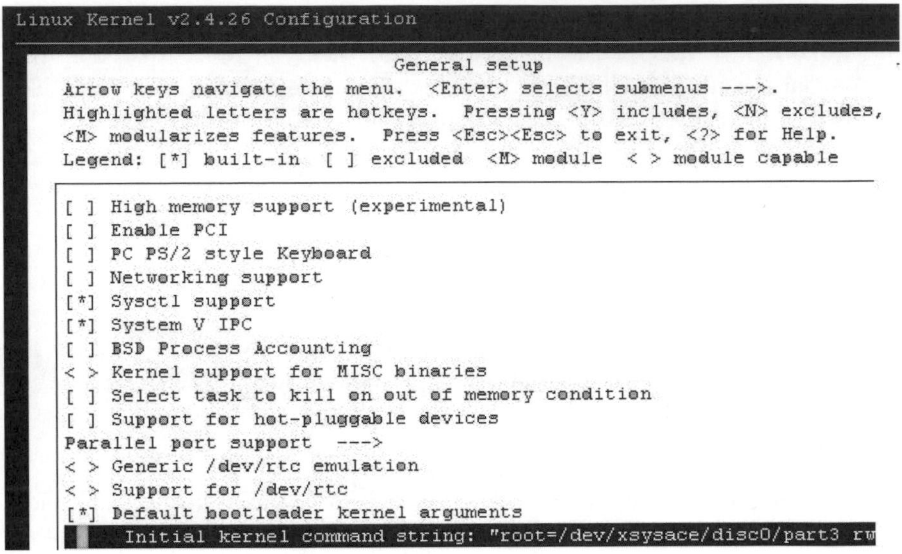

```
Linux Kernel v2.4.26 Configuration

                         General setup
 Arrow keys navigate the menu.  <Enter> selects submenus --->.
 Highlighted letters are hotkeys.  Pressing <Y> includes, <N> excludes,
 <M> modularizes features.  Press <Esc><Esc> to exit, <?> for Help.
 Legend: [*] built-in  [ ] excluded  <M> module  < > module capable

 [ ] High memory support (experimental)
 [ ] Enable PCI
 [ ] PC PS/2 style Keyboard
 [ ] Networking support
 [*] Sysctl support
 [*] System V IPC
 [ ] BSD Process Accounting
 < > Kernel support for MISC binaries
 [ ] Select task to kill on out of memory condition
 [ ] Support for hot-pluggable devices
 Parallel port support  --->
 < > Generic /dev/rtc emulation
 < > Support for /dev/rtc
 [*] Default bootloader kernel arguments
      Initial kernel command string: "root=/dev/xsysace/disc0/part3 rw
```

Figure 7.20: Kernel configuration.

4. When the version of the device driver is updated, the Makefile for the device driver must include the object file of the new device. For example, the "xsysace_sinit.c" is added to the sysace device, and thus the Makefile will be modified as follows.

   ```
   xilinx_sysace-objs += xsysace.o xsysace_sinit.o xsysace_g.o
   xsysace_intr.o xsysace_l.o
   ```

5. Use the "make dep && make zImage" command to generate the kernel image after all settings and modifications. The kernel image will be located in the directory "*arch/ppc/boot/images/*" and the file is named "**zImage.elf**."

6. The file system will be put in a flash card. First, use the partition tool, such as **fdisk**, to divide the flash card into three partitions. Partition 1 is formatted into the FAT 16 format. Here, the **mkdosfs** tool provided by Xilinx is used in the Windows OS as follows.

   `mkdosfs -s 64 -F 16 -R 1 NAME:`
   (NAME is the disk name of the flash card in the Windows OS.)

   Partition 2 and 3 are formatted in the Linux OS using the following commands.

   Partition 2 ⇒ `mkswap /dev/sda2` (use for Linux swap)
   Partition 3 ⇒ `mke2fs /dev/sda3` (use for file system)

7. After the flash card is divided into three partitions, the file system "**ramdisk.image**" provided by Xilinx is placed in Partition 3 using the following commands in the Linux OS.

   >**gzip** `-d ramdisk.image.gz`

   >**mkdir** `ramdisk_fs`

   >**mount** `-o loop ramdisk.image ramdisk_fs`

   >**cd** `ramdisk_fs`

   Finally, copy all files in the ramdisk to Partition 3 of the flash card.

8. Use the iMPACT tool to download the non-PR full bitstream, and then use XMD to download the kernel image **zImage.elf**.

◊ Write Driver for PRR

Similar to the common device driver in Linux, the PRR driver consists of the basic operations, such as open, release, write, read, and ioctl operations. Different from the software application described in Section 7.4, the control for the software accessible registers must be inserted in the basic driver operation. In the following, the basic method for inserting a new PRR driver into the Linux kernel is described.

1. Create a **xilinx_COMM** directory in the *linux_kernel/driver/char* to locate the physical driver for the standalone design.

2. The misc device method is used for the driver design. Therefore, copy the main driver source "adapter.c" from the original driver directory to modify for fitting the new PRR design. Write a Makefile to compile all driver sources.

3. Based on the interface of the partially reconfigurable template, write the read and write API functions, which are defined in the physical driver to interact with PRR. Include these APIs in the "adapter.c" source file.

```
xCOMM_write(struct file *filp, const char *buf, ...)
{
  ...
  COMM_mWriteSlaveReg2(cfg.BaseAddress, datain0);
  COMM_mWriteSlaveReg3(cfg.BaseAddress, datain1);
  ...
```

```
    COMM_mWriteSlaveReg9(cfg.BaseAddress, datain7);
    COMM_mWriteSlaveReg14(cfg.BaseAddress, cntrlin);
    ...
}
xCOMM_read(struct file *filp, char *buf, ...)
{
    ...
    dataout0 = COMM_mReadSlaveReg10(cfg.BaseAddress);
    ...
    dataout3 = COMM_mReadSlaveReg13(cfg.BaseAddress);
    cntrlout = COMM_mReadSlaveReg15(cfg.BaseAddress);
    ...
}
```

4. Open the "Config.in" file in the directory *linux_kernel/driver/char*, and insert the line "`tristate 'Xilinx COMM' CONFIG_XILINX_COMM`" into the file for adding the menu of the kernel configuration. Additionally, open the Makefile in the directory *linux_kernel/driver/char* and insert the following lines for making the kernel.

```
mod-subdirs += xilinx_gpio xilinx_ts ...  xilinx_COMM
subdir-$(CONFIG_XILINX_COMM) += xilinx_COMM
obj-$(CONFIG_XILINX_COMM) += xilinx_COMM/xilinx_COMM.o
```

5. Write a test program and compile it into an executable file. Then, use the "make dep && make zImage" command to generate the kernel image for validating the functional correctness.

◇ Integrate ICAP driver with OS

This step integrates the John Williams' ICAP driver [209] with the OS4RS for partial reconfiguration. Similar to the step for PRR driver design, first create a **xilinx_hwicap** sub-directory in the directory *linux_kernel/driver/char* to locate all ICAP sources.

1. Open the "Config.in" file in the directory *linux_kernel/driver/char*, and insert the line "`tristate 'Xilinx HWICAP' CONFIG_XILINX_HWICAP`" into the file for adding the menu of the kernel configuration.

2. Open the Makefile in the directory *linux_kernel/driver/char*, and then insert the following lines for making the kernel.

```
mod-subdirs += xilinx_gpio xilinx_ts ...  xilinx_HWICAP
subdir-$(CONFIG_XILINX_HWICAP) += xilinx_HWICAP
obj-$(CONFIG_XILINX_HWICAP) += xilinx_HWICAP/xilinx_HWICAP.o
```

3. Insert the ICAP control function into the original test program, and then compile it into an executable file.

4. Copy the software executable file and the partial bitstreams to the file system of the flash card. Then, use the "make dep && make zImage" command to generate the kernel image and execute the software application for validating functional correctness.

◇ Generate the System ACE file

This step is used to combine the static_full bitstream and the OS image file into a system ACE file.

1. Create the "xupGenace.opt" file, which includes the hardware environment information, the location of the bitstream, and the OS image file, for generating the system ACE file. The contents of the file includes:

```
-jprog
-board ml310
-hw static_full.bit
-elf zImage.elf
-ace system.ace
```

2. Select the menu item "Project" → "Launch EDK Shell." Then a command window will be displayed. Use the following command for generating the system ACE file.

```
$ xmd -tcl genace.tcl -opt xupGenace.opt
```

3. Put the "system.ace" file into Partition 1 of the flash card. Finally, power-on the ML310 board again. When the μCLinux boots successfully, run the software application to check if the partial reconfiguration can work.

If the software application can interact with the RSA and CRC designs and the partial reconfiguration can work through the ICAP, a simple OS4RS design is created.

Summary

This chapter introduces the implementation for the DPRS architecture. First, the pure hardware partial reconfigurable design for controlling the LED is illustrated. Many basic design methods and flows for partially reconfigurable system architecture are illustrated. The placement constraints for partial reconfiguration on the Xilinx Virtex families are also described. In Section 7.4, a software/hardware embedded design, which is created by the EDK, is modified into a dynamically partially reconfigurable system. Through the implementation steps, the ICAP can be used to control the partial reconfiguration without using the JTAG. Lastly, the implementation of an actual OS4RS design is described. Through the illustration of the OS4RS implementation, the PRM can be defined as a hardware task in an embedded OS. In this way, computation intensive functions can be dynamically executed as hardware tasks for enhancing the system performance. After reading this chapter, the implementation for a DPRS can be understood more concretely.

Exercises and Problems

1. What are the placement constraints for the Xilinx Virtex II Pro, Virtex 4, and Virtex 5?
2. What is the relationship between Partially Reconfigurable Region (PRR) and Partially Reconfigurable Module (PRM)?
3. What needs to be considered when multiple PRMs are to be (re)configured into the same PRR?
4. How can you modify the top-level design into a partially reconfigurable architecture?
5. What are bus macros? How many types of bus macros are there?
6. What is the difference between the driver in a traditional embedded OS and that in an OS4RS?

References

[1] L. Abeni and G. Buttazzo. Integrating multimedia applications in hard real-time systems. In *Proceedings of the 19th IEEE Real-Time Systems and Symposium*, pages 4–13. IEEE CS Press, December 1998.

[2] N. Aguirre and T. Maibaum. A temporal logic approach to the specification of reconfigurable component-based systems. In *Proceedings of the 17th IEEE International Conference on Automated Software Engineering*, pages 271–274. IEEE CS Press, September 2002.

[3] A. Ahmadinia, C. Bobda, M. Bednara, and J. Teich. A new approach for on-line placement on reconfigurable devices. In *Proceedings of the International Parallel and Distributed Processing Symposium*, volume 4, page 134. IEEE CS Press, April 2004.

[4] A. Ahmadinia, C. Bobda, and J. Teich. Online placement for dynamically reconfigurable devices. *International Journal of Embedded Systems*, 1(3–4):165–178, June 2006.

[5] A. Ahmadinia and J. Teich. Speeding up online placement for Xilinx FPGAs by reducing configuration overhead. In *Proceedings of the IFIP International Conference on VLSI-SoC*, pages 118–122, December 2003.

[6] R. Allen and D. Garlan. Formalizing architectural connection. In *Proceedings of the 16th International Conference on Software Engineering*, May 1994.

[7] E. Anderson, W. Peck, J. Stevens, J. Agron, F. Baijot, S. Warn, and D. Andrews. Supporting high-level language semantics within hardware resident threads. In *Proceedings of the 17th IEEE International Conference on Field Programmable Logic and Applications (FPL 2007)*, pages 98–103, August 2007.

[8] R. Anthony, A. Rettberg, D. Chen, I. Jahnich, G. de Boer, and C. Ekelin. Towards a dynamically reconfigurable automotive control system architecture. *Embedded System Design: Topics, Techniques and Trends*, 231:71–84, May 2007.

[9] S. R. Arikati and R. Varadarajan. A signature based approach to regularity extraction. In *Proceedings of the 1997 IEEE/ACM International Conference on Computer-Aided Design (ICCAD97)*, pages 542–545. IEEE Computer Society, 1997.

[10] J. M. Arnold, D. A. Buell, D. T. Hoang, D. V. Pryor, N. Shirazi, and M. R. Thistle. The Splash 2 Reconfigurable Processor and Applications. In *Proceedings of the International Conference on Computer Design*. CS Press, 1993.

[11] P. M. Athanas and H. F. Silverman. Processor reconfiguration through instruction-set metamorphosis. *Computer*, 26(3):11–18, Mar 1993.

[12] S. Bachl. Isomorphic subgraphs. In *Proceedings of the 7th International Symposium on Graph Drawing (GD99)*, pages 286–296. Springer-Verlag, 1999.

[13] S. Bachl and F.-J. Brandenburg. Computing and drawing isomorphic subgraphs. In *the 10th International Symposium on Graph Drawing (GD02)*, pages 74–85. Springer-Verlag, 2002.

[14] S. Bachl, F.-J. Brandenburg, and D. Gmach. Computing and drawing isomorphic subgraphs. *J. Graph Algorithms Appl.*, 8(2):215–238, 2004.

[15] S. Banerjee, E. Bozorgzadeh, and N. Dutt. Considering run-time reconfiguration overhead in task graph transformation for dynamically reconfigurable architectures. *In Proceedings of the 13th Annual IEEE Symposium on Field-Programmable Custom Computing Machines (FCCM05)*, 2005.

[16] S. Banerjee, E. Bozorgzadeh, and N. Dutt. Physically-aware hw-sw partitioning for reconfigurable architectures with partial dynamic reconfiguration. In *Proceedings of the 42st Annual Conference on Design Automation (DAC2005)*, pages 335–340, June 2005.

[17] S. Banerjee, E. Bozorgzadeh, and N. Dutt. Integrating physical constraints in hw-sw partitioning for architectures with partial dynamic reconfiguration. *In IEEE Transaction on Very Large Scale Integration Systems (T-VLSI)*, 14(11):1189–1202, November 2006.

[18] S. Banerjee, E. Bozorgzadeh, and N. Dutt. PARLGRAN: parallelism granularity selection for scheduling task chains on dynamically reconfigurable architectures. *In Proceedings of Asia and South Pacific Design Automation Conference (ASP_DAC)*, January 2006.

[19] M. Barr. A reconfigurable computing primer. *Multimedia System Design*, 9:44–47, 1998.

[20] K. Bazargan, R. Kastner, and M. Sawafzadeh. Fast template placement for reconfigurable computing systems. *IEEE Design and Test of Computers*, 17(1):68–V83, January 2000.

[21] I. Beretta. Operating system support for core management in a dynamic reconfigurable environment. Master's thesis, University of Illinois at Chicago, 2008.

[22] A. T. Berztiss. *Data Structures Theory and Practice.* Academic Press, second edition, 1975.

[23] D. C. Black and J. Donovan. *SystemC: From the Ground Up.* Kluwer Academic Publishers, 2004.

[24] B. Blodget, P. James-Roxby, E. Keller, S. McMillan, and P. Sundararajan. A self-reconfiguring platform. In *Proceedings of the 13th IEEE International Conference on Field Programmable Logic and Application (FPL 2003)*, volume 2778 of *Lecture Notes in Computer Science.* Springer, 2003.

[25] C. Bobda, A. Ahmadinia, K. Rajesham, M. Majer, and A. Niyonkuru. Partial configuration design and implementation challenges on Xilinx Virtex FPGAs. In *ARCS Workshops*, pages 61–66, 2005.

[26] M. Borgatti, A. Fedeli, U. Rossi, J.-L. Lambert, I. Moussa, F. Fummi, C. Marconcini, and G. Pravadelli. A verification methodology for reconfigurable systems. In *Proceedings of the 5th International Workshop on Microprocessor Test and Verification (MTV)*, pages 85–90. IEEE CS Press, September 2004.

[27] L. Bowen. Introduction to contemporary mathematics. In *http://www.ctl.ua.edu/math103/scheduling/scheduling_algorithms.htm.*

[28] G. Brebner. A virtual hardware operating system for the Xilinx XC6200. In *Proceedings of the 6th International Workshop on Field-Programmable Logic and Applications*, pages 327–336. Springer-Verlag, September 1996.

[29] G. Brebner and O. Diessel. Chip-based reconfigurable task management. In *Proceedings of the 11th IEEE International Conference on Field-Programmable Logic and Applications (FPL 2001)*, pages 182–191. Springer-Verlag, August 2001.

[30] P. Butel, G. Habay, and A. Rachet. Managing partial dynamic reconfiguration in Virtex-II Pro FPGAs. *Xcell Journal Online*, 2004.

[31] T.J. Callahan, J.R. Hauser, and J. Wawrzynek. The garp architecture and c compiler. *Computer*, 33(4):62–69, April 2000.

[32] F. Cancar, M. D. Santambrogio, and D. Sciuto. A design flow tailored for self dynamic reconfigurable architecture. In *IEEE International Parallel and Distributed Processing Symposium (IPDPS08) - Reconfigurable Architecture Workshop - RAW*, page available online, 2008.

[33] J. M. P. Cardoso. Loop dissevering: a technique for temporally partitioning loops in dynamically reconfigurable computing platforms. In *Proceedings of the International Parallel and Distributed Processing Symposium (IPDPS03)*, page 181.2. IEEE Computer Society, 2003.

[34] J. M. P. Cardoso. On combining temporal partitioning and sharing of functional units in compilation for reconfigurable architectures. *IEEE Trans. Computers*, 52(10):1362–1375, 2003.

[35] J. M. P. Cardoso and H. C. Neto. Fast hardware compilation of behaviors into an FPGA-based dynamic reconfigurable computing system. In *Proceedings of the XII Symposium on Integrated Circuits and Systems Design (SBCCI99)*, pages 150–153, October 1999.

[36] J. M. P. Cardoso and H. C. Neto. Macro-based hardware compilation of Java bytecodes into a dynamic reconfigurable computing system. In *Proceedings of the IEEE Workshop on FPGAs for Custom Computing Machines*, pages 2–11. IEEE, 1999.

[37] J. M. P. Cardoso and H. C. Neto. Compilation for FPGA-based reconfigurable hardware. *IEEE Design & Test of Computers*, 20(2):65–75, 2003.

[38] J. M. P. Cardoso and H. C. Neto. Compilation for FPGA-Based Reconfigurable Hardware. *IEEE Design & Test of Computers*, 20(2):65–75, 2003.

[39] E. Carvalho, N. Calazans, E. Bri, and F. Moraes. PaDReH: a framework for the design and implementation of dynamically and partially reconfigurable systems. In *Proceedings of the 17th Symposium on Integrated Circuits and System Design (SBCCI)*, pages 10–15. ACM Press, 2004.

[40] G. Chaitin. Register allocation and spilling via graph coloring. *SIGPLAN Not.*, 39(4):66–74, 2004.

[41] G. J. Chaitin, M. A. Auslander, A. K. Chandra, J. Cocke, M. E. Hopkins, and P. W. Markstein. Register allocation via coloring. In *Computer Languages*, volume 6, pages 47–57, 1981.

[42] Y.-H. Chen and P.-A. Hsiung. Hardware task scheduling and placement in operating systems for dynamically reconfigurable soc. In *Proceedings of the International Conference on Embedded and Ubiquitous Computing (EUC)*, volume 3824, pages 489–498. Springer-Verlag, December 2005.

[43] C.-C. Chiang. Hardware/software real-time relocatable task scheduling and placement in dynamically partial reconfigurable systems. Master's thesis, National Chung Cheng University, June 2007.

[44] A. Chowdary, S. Kale, P. K. Saripella, N. K. Sehgal, and R. K. Gupta. Extraction of functional regularity in datapath circuits. *IEEE Transactions on Computer-Aided Design of Integrated Circuits and Systems*, 18(9):1279–1296, 1999.

[45] K. Compton and S. Hauck. Reconfigurable computing: a survey of systems and software. *ACM Computing Surveys*, 34(2):171–210, 2002.

[46] S. A. Cook. The complexity of theorem-proving procedures. In *Proceedings of the 3rd ACM Symposium on Theory of Computing*, pages 151–158, 1971.

[47] S. Corbetta, F. Ferrandi, M. Morandi, M. Novati, M. D. Santambrogio, and D. Sciuto. Two novel approaches to online partial bitstream relocation in a dynamically reconfigurable system. In *IEEE Computer Society Annual Symposium on VLSI*, pages 457–458, May 2007.

[48] IBM Corporation. *The CoreConnect Bus Architecture, white paper.* International Business Machines Corporation, 2004.

[49] K. Danne and M. Platzner. Periodic real-time scheduling for FPGA computers. In *Proceedings of the 3rd International Workshop on Intelligent Solutions in Embedded Systems*, pages 117–127. IEEE CS Press, May 2005.

[50] M. Davis, G. Logemann, and D. Loveland. A machine program for theorem proving. *Communications of the ACM*, 5:394–397, 1962.

[51] M. Davis and H. Putnam. A computing procedure for quantification theory. *Journal of the ACM*, 7(3):201–215, 1960.

[52] A. DeHon, J. Adams, M. DeLorimier, N. Kapre, Y. Matsuda, H. Naeimi, M. Vanier, and M. Wrighton. Design patterns for reconfigurable computing. In *Proceedings of the 12th Annual IEEE Symposium on Field-Programmable Custom Computing Machines (FCCM04)*, pages 13–23, April 2004.

[53] A. E. Dehon. DPGA-coupled microprocessors: commodity ICs for the early 21st century. In *Proceedings of the IEEE Workshop on FPGAs for Custom Computing Machines*, pages 31–39, 1994.

[54] R. P. Dick, D. L. Rhodes, and W. Wolf. TGFF: task graphs for free. In *Proceedings of the 6th International Workshop on Hardware/Software Codesign*, pages 97–101. IEEE CS Press, March 1998.

[55] O. Diessel, H. Elgindy, M. Middendorf, H. Schmeck, and B. Schmidt. Dynamic scheduling of tasks on partially reconfigurable FPGAs. In *IEE Proceedings on Computers and Digital Techniques*, pages 181–188, May 2000.

[56] A. Donato, F. Ferrandi, M. Redaelli, M. D. Santambrogio, and D. Sciuto. Caronte: a complete methodology for the implementation of partially dynamically self-reconfiguring systems on FPGA platforms. In *Proceedings of the 13th Annual IEEE Symposium on Field-Programmable Custom Computing Machines (FCCM05)*, pages 321–322, April 2005.

[57] A. Donato, F. Ferrandi, M. Redaelli, M. D. Santambrogio, and D. Sciuto. Exploiting partial dynamic reconfiguration for SoC design of complex application on FPGA platforms. *VLSI-SoC: From Systems to Silicon*, pages 87–109, 2007.

[58] A. Donato, F. Ferrandi, M. D. Santambrogio, and D. Sciuto. Operating system support for dynamically reconfigurable SoC architectures. In *Proceedings of the IEEE International SOC Conference*, pages 233–238, September 2005.

[59] M. Dyer and M. Wirz. *Reconfigurable System on FPGA*. Diploma Thesis, ETH Zurich, Winter Term, 2001/2002.

[60] V. Ech, P. Kalra, R. LeBlanc, and J. McManus. In-Circuit Partial Reconfiguration of RocketIO Attributes. Technical Report XAPP662, Xilinx Inc., January 2003.

[61] R. Enzler, C. Plessi, and M. Platzner. System-level performance evaluation of reconfigurable processors. *Microprocessors and Microsystems*, 29(2–3):63–73, April 2005.

[62] G. Estrin. Organization of computer systems: The fixed plus variable structure computer. In *Proceedings of the Western Joint Computer Conference*, pages 33–40, May 1960.

[63] J. Resano et al. Specific scheduling support to minimize the reconfiguration overhead of dynamically reconfigurable hardware. In *Proceedings of the 41st Annual Conference on Design Automation (DAC2004)*, pages 119–125, June 2004.

[64] F. Ferrandi, L. Fossati, M. Lattuada, G. Palermo, D. Sciuto, and A. Tumeo. Partitioning and mapping for the hArtes European project, 2007.

[65] F. Ferrandi, M. Morandi, M. Novati, M. D. Santambrogio, and D. Sciuto. Dynamic reconfiguration: Core relocation via partial bitstreams filtering with minimal overhead. In *8th International Symposium on System-on-Chip*, pages 33–36, November 2006.

[66] F. Ferrandi, M. Redaelli, M. D. Santambrogio, and D. Sciuto. Solving the coloring problem to schedule on partially dynamically reconfigurable hardware. In *13th International Conference on Very Large Scale Integration*, pages 97–102, 2005.

[67] G. Fey, R. Drechsler, and M. Ali. An approach to formal verification of reconfigurable systems. In *Proceedings of the 1st IFIP WG 10.5 Workshop on Frontiers in Automotive Electronics*, 2003.

[68] R. J. Fong, S. J. Harper, and P. M. Athanas. A versatile framework for FPGA field updates: an application of partial self-reconfiguation. *rsp*, 00:117, 2003.

[69] W. Fornaciari and V. Piuri. Virtual FPGAs: some steps behind the physical barriers. In *IPPS/SPDP Workshops*, pages 7–12, 1998.

[70] S. Ganesan and R. Vemuri. An integrated temporal partitioning and partial reconfiguration technique for design latency improvement. In *Proceedings of the Conference on Design, Automation and Test in Europe (DATE00)*, pages 27–30, 2000.

[71] S. Ghiasi and M. Sarrafzadeh. Optimal reconfiguration sequence management. *In Proceedings of Asia South Pacific Design Automation Conference (ASP_DAC)*, pages 359–365, 2003.

[72] M. Giorgetta, M. D. Santambrogio, P. Spoletini, and D. Sciuto. A graph-coloring approach to the allocation and tasks scheduling for reconfigurable architectures. In *14th IFIP International Conference on Very Large Scale Integration - IFIP VLSI-SOC*, pages 24–29, October 2006.

[73] M. Gokhale and P. S. Graham. *Reconfigurable Computing – Accelerating Computation with Field-Programmable Gate Arrays.* Springer, 2005.

[74] S. C. Goldstein, H. Schmit, M. Budiu, S. Cadambi, M. Moe, and R. R. Taylor. Piperench: a reconfigurable architecture and compiler. *Computer*, 33(4):70–77, Apr 2000.

[75] J. Gu, P. W. Purdom, J. Franco, and B. W. Wah. Algorithms for the satisfiability (SAT) problem: a survey. *DIMACS Series in Discrete Math and Theoretical Computer Science*, 35:19–151, 1997.

[76] J. Hagemeyer, B. Keltelhoit, M. Koester, and M. Pomnann. A design methodology for communication infrastructures on partially reconfigurable FPGAs. In *Proceedings of the 17th IEEE International Conference on Field Programmable Logic and Applications (FPL 2007)*, pages 331–338, August 2007.

[77] J. Hagemeyer, B. Kettelhoit, M. Koester, and M. Porrmann. Design of homogeneous communication infrastructures for partially reconfigurable FPGAs. In *Proceedings of the International Conference on Engineering of Reconfigurable Systems and Algorithms*, pages 238–247. CSREA Press, June 2007.

[78] M. Handa, R. Radhakrishnan, M. Mukherjee, and R. Vemuri. A fast macro based compilation methodology for partially reconfigurable FPGA designs. In *16th International Conference on VLSI Design*, pages 91–96, January 2003.

[79] M. Handa and R. Vemuri. An efficient algorithm for finding empty rectangles for online FPGA placement. In *Proceedings of the 41st Annual Conference on Design Automation (DAC2004)*, pages 960–V965. ACM Press, June 2004.

[80] M. Handa and R. Vemuri. Hardware assisted two dimensional ultra fast online placement. *International Journal of Embedded Systems (IJES)*, 1:291–V299, June 2006.

[81] R. Hartenstein. Why we need reconfigurable computing education. In *Proceedings of the 1st International Workshop on Reconfigurable Computing Education*, pages 1–11, March 2006.

[82] J.R. Hauser and J. Wawrzynek. Garp: a MIPS processor with a reconfigurable coprocessor. In *Proceedings of the 5th Annual IEEE Symposium on FPGAs for Custom Computing Machines*, pages 12–21, Apr 1997.

[83] R. Hecht, S. Kubisch, H. Michelsen, E. Zeeb, and D. Timmermann. A distributed object system approach for dynamic reconfiguration. In *Proceedings of the 20th International Symposium on Parallel and Distributed Processing Symposium*, April 2006.

[84] J. L. Hennessy and D. A. Patterson. *Computer Architecture - A Quantitative Approach (Fourth Edition)*. Morgan Kaufmann, 2007.

[85] E. L. Horta and J. W. Lockwood. PARBIT: a tool to transform bitfiles to implement partial reconfiguration of field programmable gate arrays (FPGAs). *Washington University, Department of Computer Science, Technical Report WUCS-01-13*, July 2001.

[86] E. L. Horta and J. W. Lockwood. Automated method to generate bitstream intellectual property cores for Virtex FPGAs. In *Proceedings of the International Conference on Field Programmable Logic and Application*, pages 975–979, 2004.

[87] E. L. Horta, J. W. Lockwood, D. E. Taylor, and D. Parlour. Dynamic hardware plugins in an FPGA with partial run-time reconfiguration. In *Proceedings of the 39st Annual Conference on Design Automation (DAC2002)*, pages 343–348. ACM, 2002.

[88] P.-A. Hsiung, C.-H. Huang, and Y.-H. Chen. Hardware task scheduling and placement in operating systems for dynamically reconfigurable SoC. *Journal of Embedded Computing*, 2008. To appear.

[89] P.-A. Hsiung, C.-H. Huang, and C.-F. Liao. Perfecto: A SystemC-based performance evaluation framework for dynamically partially reconfigurable systems. In *Proceedings of the 16th IEEE International Conference on Field Programmable Logic and Applications (FPL 2006)*, pages 190–198. IEEE CS Press, August 2006.

[90] P.-A. Hsiung, C.-S. Lin, and C.-F. Liao. Perfecto: A SystemC-based design space exploration framework for dynamically partially reconfigurable systems. *ACM Transactions on Reconfigurable Technology and Systems*, 1(3):1–30, 2008.

[91] P.-A. Hsiung and C.-W. Liu. Exploiting hardware and software low power techniques for energy efficient co-scheduling in dynamically reconfigurable systems. In *Proceedings of the 17th IEEE International Conference on Field-Programmable Logic and Applications (FPL 2007)*, pages 165–170. IEEE CS Press, August 2007.

[92] P.-A. Hsiung, P.-H. Lu, and C.-W. Liu. Energy efficient hardware-software co-scheduling in dynamically reconfigurable systems. In *Proceedings of the International Conference on Hardware-Software Codesign and System Synthesis*, pages 87–92. ACM Press, September 2007.

[93] C.-H. Huang and P.-A. Hsiung. UML-based hardware/software co-design for partially reconfigurable systems. In *Proceedings of the 13th IEEE Asia-Pacific Computer Systems Architecture Conference (ACSAC)*, pages 1–6, August 2008. doi: 10.1109/APCSAC.2008.4625436.

[94] R. D. Hudson, D. Lehn, J. Hess, J. Atwell, D. Moye, K. Shiring, and P. Athanas. Spatio-temporal partitioning of computational structures onto configurable computing machines. In *Configurable Computing: Technology and Applications, Proc. SPIE 3526*, pages 62–71. SPIE – The International Society for Optical Engineering, 1998.

[95] Xilinx Inc. Two Flows of Partial Reconfiguration: Module Based or Difference Based. Technical Report XAPP290, Xilinx Inc., November 2003.

[96] Xilinx Inc. Virtex Series Configuration Architecture User Guide. Technical Report XAPP151, Xilinx Inc., March 2003.

[97] Xilinx Inc. Opb hwicap (v1.00.b) product specification. Technical report, Xilinx Inc., March 2005.

[98] Xilinx Inc. *Early Access Partial Reconfiguration Guide.* Xilinx Inc., 2006.

[99] Xilinx Inc. *Embedded Development Kit EDK 8.2i.* Xilinx Inc., 2006.

[100] Xilinx Inc. Virtex-4 configuration user guide. Technical Report ug71, Xilinx Inc., January 2007.

[101] Xilinx Inc. Virtex-4 user guide. Technical Report ug70, Xilinx Inc., March 2007.

[102] Xilinx Inc. Virtex-5 user guide. Technical Report ug190, Xilinx Inc., February 2007.

[103] Xilinx Inc. Xilinx : Virtex series FPGAs. http://www.xilinx.com/products/silicon_solutions/fpgas/virtex/index.htm, 2007.

[104] Xilinx Inc. *PlanAhead User Guide.* Xilinx Inc., July 27, 2007.

[105] K. Jafri, N. Jafri, and S. Khan. Constraint based temporal partitioning model for partial reconfigurable architectures. *In Proceedings of IEEE INIMIC*, pages 242–246, 2003.

[106] TU Kaiserslautern. The configware page, 2007. http://configware.org.

[107] TU Kaiserslautern. The flowware page, 2007. http://flowware.net.

[108] H. Kalte and M. Porrmann. REPLICA2Pro: Task relocation by bitstream manipulation in Virtex-II/Pro FPGAs. In *Proc. of the ACM International Conference on Computing Frontiers*, 2006.

[109] H. Kalte, M. Porrmann, and U. Ruckert. A prototyping platform for dynamically reconfigurable system on chip designs. In *Proceedings of the IEEE Workshop Heterogeneous Reconfigurable Systems-on-Chip*, 2002.

[110] H. Kalte, M. Porrmann, and U. Rückert. System-on-programmable-chip approach enabling online fine-grained 1D-placement. In *Proceedings of the 18th International Parallel and Distributed Processing Symposium (IPDPS) - Reconfigurable Architectures Workshop (RAW)*. IEEE Computer Society, 2004.

[111] R. K. Karanam, A. Ravindran, A. Mukherjee, C. Gibas, and A. B. Wilkinson. Using FPGA-based hybrid computers for bioinformatics applications. *Xcell journal*, 2006.

[112] M. Kaul and R. Vemuri. Optimal temporal partitioning and synthesis for reconfigurable architectures. In *Proceedings of the Conference on Design, Automation and Test in Europe (DATE98)*, February 1998.

[113] M. Kaul and R. Vemuri. Temporal partitioning combined with design space exploration for latency minimization of run-time reconfigured design. In *Proceedings of the Conference on Design, Automation and Test in Europe (DATE99)*, pages 202–209, 1999.

[114] M. Kaul, R. Vemuri, S. Govindarqjan, and I. Ouaiss. An automated temporal partitioning and loop fission approach for FPGA based reconfigurable synthesis of dsp applications. In *Proceedings of the 36st Annual Conference on Design Automation (DAC1999)*, pages 616–622. IEEE Computer Society, 1999.

[115] D. Koch, C. Beckhoff, and J. Teich. RECOBUS-Builder - a novel tool and technique to build statically and dynamically reconfigurable systems for FPGAs. In *Proceedings of the 18th IEEE International Conference on Field Programmable Logic and Applications (FPL 2008)*, pages 119–124, September 2008.

[116] D. Koch, C. Haubelt, and J. Teich. Efficient reconfigurable on-chip buses for FPGAs. In *Proceedings of the 16th Annual IEEE Symposium on Field-Programmable Custom Computing Machines (FCCM08)*, 2008.

[117] M. Koester, H. Kalte, and M. Porrmann. Task placement for heterogeneous reconfigurable architectures. In *Proceedings of the IEEE 2005 Conference on Field-Programmable Technology (FPT05)*, pages 343–348, December 2005.

[118] Y. E. Krasteva, E. de la Torre, T. Riesgo, and D. Joly. Virtex II FPGA bitstream maniplation: Application to reconfiguration control systems. In *Proceedings of the 16th IEEE International Conference on Field Programmable Logic and Applications (FPL 2006)*, pages 1–4, August 2006.

[119] Y. E. Krasteva, A. B. Jimeno, E. de la Torre, and T. Riesgo. Straight method for reallocation of complex cores by dynamic reconfiguration in Virtex II FPGAs. In *16th IEEE International Workshop on Rapid System Prototyping, (RSP2005)*, June 2005.

[120] J. Krinke. Identifying similar code with program dependence graphs. In *Proceedings of Eighth Working Conference on Reverse Engineering*, pages 301–309, 2001.

[121] T. Kutzschebauch. Efficient logic optimization using regularity extraction. In *2000 IEEE International Conference on Computer Design (ICCD00)*, pages 487–493, 2000.

[122] D. Lavenier, S. Guyetant, S. Derrien, and S. Rubini. A reconfigurable parallel disk system for filtering genomic banks. In *Proceedings of the International Conference on Engineering of Reconfigurable Systems and Algorithms*, 2003.

[123] M.-H. Lee, H. Singh, G. Lu, N. Bagherzadeh, F. J. Kurdahi, E. M. C. Filho, and V. C. Alves. Design and implementation of the morphosys reconfigurable computing processor. *J. VLSI Signal Process. Syst.*, 24(2–3):147–164, 2000.

[124] H.-W. Liao. Multi-objective placement of reconfigurable hardware tasks in real-time systems. Master's thesis, National Chung Cheng University, June 2007.

[125] X. P. Ling and H. Amano. Performance evaluation of wasmii: a data driven computer on a virtual hardware. In *PARLE*, pages 610–621, 1993.

[126] C.-W. Liu. Energy efficient hardware/software co-scheduling in reconfigurable systems. Master's thesis, National Chung Cheng University, 2006.

[127] S. M. Loo and B. E. Wells. Task scheduling in a finite-resource, reconfigurable hardware/software codesign environment. *INFORMS Journal on Computing*, 18(2):151–172, Spring 2006.

[128] G. Lu, H. Singh, M.-H. Lee, N. Bagherzadeh, F. J. Kurdahi, and E. M. C. Filho. The morphosys parallel reconfigurable system. In *European Conference on Parallel Processing*, pages 727–734, 1999.

[129] P. Lysaght, B. Blodget, J. Mason, J. Young, and B. Bridgford. Invited paper: Enhanced architectures, design methodologies and CAD tools for dynamic reconfiguration of Xilinx FPGAs. In *Proceedings of the 16th IEEE International Conference on Field Programmable Logic and Applications (FPL 2006)*, pages 1–6. IEEE CS Press, August 2006.

[130] R. Maestre, M. Fernández, R. Hermida, and N. Bagherzadeh. A framework for scheduling and context allocation in reconfigurable computing. In *ISSS*, pages 134–140, 1999.

[131] T. Marescaux, J.-Y. Mignolet, A. Bartic, W. Moffat, D. Verkest, S. Vernalde, and R. Lauwereins. Networks on chip as hardware components of an OS for reconfigurable systems. In *Proceedings of the 13th IEEE International Conference on Field Programmable Logic and Applications (FPL 2003)*, volume 2778 of *Lecture Notes in Computer Science (LNCS)*, pages 595–605, September 2003.

[132] B. Mei, P. Schaumont, and S. Vernalde. A hardware-software paritioning and scheduling algorithm for dynamically reconfigurable embedded systems. In *Proceedings of the 11th ProRISC Workshop on Circuits, Systems and Signal Processing*, November 2000.

[133] J. Mignolet, S. Vernalde, D. Verkest, and R. Lauwereins. Enabling hardware-software multitasking on a reconfigurable computing platform for networked portable multimedia appliances. In *Proceedings of the International Conference on Engineering of Reconfigurable Systems and Algorithms*, pages 116–122. IEEE CS Press, June 2002.

[134] J.-Y. Mignolet, V. Nollet, P. Coene, D. Verkest, S. Vernalde, and R. Lauwereins. Infrastructure for design and management of relocatable tasks in a heterogeneous reconfigurable system-on-chip. In *Proceedings of the Conference on Design, Automation and Test in Europe (DATE03)*, pages 986–991, March 2003.

[135] M. Murgida, A. Panella, V. Rana, M. D. Santambrogio, and D. Sciuto. Fast IP-core generation in a partial dynamic reconfiguration workflow. In *14th IFIP International Conference on Very Large Scale Integration - IFIP VLSI-SOC*, pages 74–79, October 2006.

[136] J. Noguera and R. M. Badia. Dynamic run-time HW/SW scheduling techniques for reconfigurable architectures. In *Proceedings of the International Conference on Hardware-Software Codesign*, pages 205–210. ACM Press, May 2002.

[137] J. Noguera and R. M. Badia. System-level power-performance trade-offs in task scheduling for dynamically reconfigurable architectures. In *Pro-*

ceedings of the International Conference on Compilers, Architectures, and Synthesis for Embedded Systems, pages 78–83. ACM Press, October 2003.

[138] J. Noguera and R. M. Badia. Multitasking on reconfigurable architectures: Microarchitecture support and dynamic scheduling. *ACM Transactions on Embedded Computing Systems*, 3(2):385–406, May 2004.

[139] V. Nollet, P. Coene, D. Verkest, S. Vernalde, and R. Lauwereins. Designing an operating system for a heterogeneous reconfigurable SoC. In *Proceedings of the 17th International Symposium on Parallel and Distributed Processing*, page 174. IEEE CS Press, April 2003.

[140] V. Nollet, T. Marescaux, D. Verkest, J.-Y. Mignolet, and S. Vernalde. Operating system controlled network-on-chip. In *Proceedings of the 41st Annual Conference on Design Automation (DAC2004)*, pages 256–259, June 2004.

[141] M. J. Osborne. *An Introduction to Game Theory.* Oxford University Press, 2002.

[142] Open SystemC Initiative (OSCI). *SystemC User's Guide.* http://www.systemc.org/, 2008.

[143] J. Ou, S. Choi, and V. K. Prasanna. Energy-efficient hardware/software co-synthesis for a class of applications on reconfigurable SoCs. *International Journal of Embedded Systems*, 1(1/2):91–102, January 2005.

[144] I. Ouaiss, S. Govindarajan, V. Srinivasan, M. Kaul, and R. Vemuri. An integrated partitioning and synthesis system for dynamically reconfigurable multi-FPGA architectures. In *IPPS/SPDP Workshops*, pages 31–36, 1998.

[145] A. Pandey and R. Vemuri. Combined temporal partitioning and scheduling for reconfigurable architectures. In *Reconfigurable Technology: FPGAs for Computing and Applications, Proc. SPIE 3844*, pages 93–103, Bellingham, WA, 1999. SPIE – The International Society for Optical Engineering.

[146] H. Parizzi, A. Niktash, N. Bagherzadeh, and F.J. Kurdahi. MorphoSys: A Coarse Grain Reconfigurable Architecture for Multimedia Applications. In *Euro-Par 2002*, 2002.

[147] A. Pelkonen, K. Masselos, and M. Cupák. System-level modeling of dynamically reconfigurable hardware with SystemC. In *Proceedings of the 10th Reconfigurable Architectures Workshop (RAW)*, page 174. IEEE CS Press, April 2003.

[148] A. Pelkonen, K. Masselos, and M. Cupék. System-level modeling of dynamically reconfigurable hardware with systemc. In *Proceedings of the*

17th Int. Symposium on Parallel and Distributed Processing (PDPS03). IEEE Computer Society, 2003.

[149] R. Pellizzoni and M. Caccamo. Adaptive allocation of software and hardware real-time tasks for FPGA-based embedded systems. In *Proceedings of the 12th IEEE Real-Time and Embedded Technology and Applications Symposium (RTAS)*, pages 208–220. IEEE CS Press, April 2006.

[150] K. M. G. Purna and D. Bhatia. Temporal partitioning and scheduling data flow graphs for reconfigurable computers. *IEEE Trans. Comput.*, 48(6):579–590, 1999.

[151] Y. Qu, K. Tiensyrjä, and K. Masselos. System-level modeling of dynamically reconfigurable co-processors. In *Proceedings of the 14th IEEE International Conference on Field Programmable Logic and Application (FPL 2004)*, volume 3203, pages 881–885. Springer Verlag, August 2004.

[152] V. Rana, M. D. Santambrogio, and D. Sciuto. Dynamic reconfigurability in embedded system design. In *IEEE International Symposium on Circuits and Systems*, pages 2734–2737. IEEE, May 2007.

[153] D.S. Rao and F.J. Kurdahi. On clustering for maximal regularity extraction. *IEEE Transactions on Computer-Aided Design of Integrated Circuits and Systems*, 12(8):1198–1208, 1993.

[154] J. Resano, D. Mozos, and F. Catthoor. A hybrid prefetch scheduling heuristic to minimize at run-time the reconfiguration overhead of dynamically reconfigurable hardware. In *Proceedings of the Conference on Design, Automation and Test in Europe (DATE05)*, volume 1, pages 106–111. IEEE CS Press, March 2005.

[155] J. Resano, D. Mozos, D. Verkest, F. Catthoor, and S. Vernalde. Specific scheduling support to minimize the reconfiguration overhead of dynamically reconfigurable hardware. In *Proceedings of the 41st Annual Conference on Design Automation (DAC2004)*, pages 119–124. ACM Press, June 2004.

[156] J. Resano, D. Mozos, D. Verkest, S. Vernalde, and F. Catthoor. Run-time minimization of reconfiguration overhead in dynamically reconfigurable systems. In *Proceedings of the 13th IEEE International Conference on Field-Programmable Logic and Applications (FPL 2003)*, pages 585–594. Springer Verlag, September 2003.

[157] Tanner Research. Reconfigurable computing, December 2007. http://www.reconfig-computing.com/default.htm.

[158] T. Rissa, A. Donlin, and W. Luk. Evaluation of SystemC modelling of reconfigurable embedded systems. In *Proceedings of the Conference on*

Design, Automation and Test in Europe (DATE05), volume 3, pages 253–258, March 2005.

[159] A.P.E. Rosiello, F. Ferrandi, D. Pandini, and D. Sciuto. A hash-based approach for functional regularity extraction during logic synthesis. *IEEE Computer Society Annual Symposium on VLSI (ISVLSI07)*, pages 92–97, March 2007.

[160] P.J. Roxby, E.C. Prada, and S. Charlwood. Core-based design methodology for reconfigurable computing applications. *IEEE Proceeding Comp. Digital Techonology*, 157(3), May 2000.

[161] M. Sanchez-Elez, M. Fernandez, M. Anido, H. Du, N. Bagherzadeh, and R. Hermida. Low-energy data management for different on-chip memory levels in multi-context reconfigurable architectures. In *Proceedings of the Conference on Design, Automation and Test in Europe (DATE03)*, volume 1, pages 36–41. IEEE CS Press, March 2003.

[162] M. Santambrogio. *Hardware-Software Codesign Methodologies for Dynamically Reconfigurable Systems.* PhD thesis, Politecnico Di Milano, Italy, February 2008.

[163] M. D. Santambrogio and D. Sciuto. Design methodology for partial dynamic reconfiguration: a new degree of freedom in the hw/sw codesign. In *IEEE International Parallel and Distributed Processing Symposium (IPDPS08) - Reconfigurable Architecture Workshop - RAW*, page available online, 2008.

[164] J. Schewel. Hardware/software codesign system using reconfigurable computing technology. In *Proceedings of the 12th International Parallel Processing Symposium and the 9th Symposium on Parallel and Distributed Processing (IPPS/SPDP)*, pages 620–625, March 1998.

[165] P. Sedcole, B. Blodget, T. Becker, J. Anderson, and P. Lysaght. Modular dynamic reconfiguration in Virtex FPGAs. *IEE Proceedings Computers and Digital Techniques*, 153(3):157–164, 2006.

[166] L. Shang, R. P. Dick, and N. K. Jha. SLOPES: Hardware-software cosynthesis of low-power real-time distributed embedded systems with dynamically reconfigurable FPGAs. *IEEE Transactions on Computer-Aided Design of Integrated Circuits and Systems*, 26(3):508–526, March 2007.

[167] L. Shang and N. Jha. Hardware-software co-synthesis of low power real-time distributed embedded systems with dynamically reconfigurable FPGAs. In *Proceedings of the International Conference on VLSI Design*, pages 345–352. IEEE CS Press, January 2002.

[168] J.-S. Shen and P.-A. Hsiung. *Reconfigurable Network-on-Chip.* IGI Global, USA, 2009.

[169] K.-J. Shih, P.-A. Hsiung, and C.-C. Hung. Reconfigurable hardware module sequencer – a tradeoff between networked and data flow architectures. In *Proceedings of the International Conference on Field-Programmable Technology (ICFPT)*, pages 237–240. IEEE Computer Society Press, December 2007.

[170] H. Singh, M.-H. Lee, G. Lu, F.J. Kurdahi, N. Bagherzadeh, and E. M. C. Filho. MorphoSys: an integrated reconfigurable system for data-parallel and computation-intensive applications. *IEEE Transactions on Computers*, 49(5):465–481, 2000.

[171] H. Singh, G. Lu, M.-H. Lee, E. Filho, R. Maestre, F. Kurdahi, and N. Bagherzadeh. Morphosys: case study of a reconfigurable computing system targeting multimedia applications. In *Proceedings of the 37th Annual Conference on Design Automation (DAC2000)*, pages 573–578, 2000.

[172] S. Singh and C. J. Lillieroth. Formal verification of reconfigurable cores. In *Proceedings of the 7th IEEE Symposium on Field-Programmable Custom Computing Machines (FCCM)*, pages 25–32. IEEE CS Press, April 1999.

[173] L. Singhal and E. Bozorgzadeh. Multi-layer floorplanning on a sequence of reconfigurable design. In *Proceedings of the 16th IEEE International Conference on Field Programmable Logic and Applications (FPL 2006)*, pages 1–8, 2006.

[174] I. Skilarova and A. Ferrari. Reconfigurable hardware SAT solvers: A survey of systems. *IEEE Transactions on Computers*, 53(11):1449–1461, November 2004.

[175] H. K.-H. So and R. W. Brodersen. Improving usability of FPGA-based reconfigurable computers through operating system support. In *Proceedings of the 16th IEEE International Conference on Field Programmable Logic and Applications (FPL 2006)*, pages 349–354, August 2006.

[176] H. K.-H. So and R. W. Brodersen. A unified hardware/software runtime environment for FPGA-based reconfigurable computers using BORPH. *ACM Transactions on Embedded Computing Systems*, 7(2):(article no. 14), February 2008.

[177] C. Steiger, H. Walder, and M. Platzner. Heuristics for online scheduling real-time tasks to partially reconfigurable devices. In *Proceedings of the 13th IEEE International Conference on Field Programmable Logic and Applications (FPL 2003)*, pages 575–584, September 2003.

[178] C. Steiger, H. Walder, and M. Platzner. Operating systems for reconfigurable embedded platforms: Online scheduling of real-time tasks. *IEEE Transactions on Computers*, 53(11):1393–1407, November 2004.

[179] C. Steiger, H. Walder, M. Platzner, and L. Thiele. Online scheduling and placement of real-time tasks to partially reconfigurable devices. In *Proceedings of the 24th IEEE International Real-Time Systems Symposium*, page 224. IEEE CS Press, December 2003.

[180] H.-Y. Sun. Dynamic hardware-software task switching and relocation mechanisms for reconfigurable systems. Master's thesis, National Chung Cheng University, July 2007.

[181] J. Tabero, J. Septien, H. Mecha, and D. Mozos. A low fragmentation heuristic for task placement in 2D RTR HW management. In *Proceedings of the International Conference on Field-Programmable Logic and Applications*, pages 241–V250. Springer Verlag, August 2004.

[182] J. Tabero, J. Septien, H. Mecha, and D. Mozos. Task placement heuristic based on 3D-adjacency and look-ahead in reconfigurable systems. In *Proceedings of the Asia South Pacific Design Automation Conference (ASP_DAC)*, pages 24–V27. ACM Press, January 2006.

[183] S. Tapp. Configuration quick start guidelines. *XAPP501*, July 2003.

[184] D. E. Taylor, J. W. Lockwood, and S. Dharmapurikar. Generalized rad module interface specification of the field programmable port extender (fpx). *Washington University, Department of Computer Science. Version 2.0, Technical Report*, January 2000.

[185] M. B. Taylor, J. Kim, J. Miller, D. Wentzlaff, F. Ghodrat, B. Greenwald, H. Hoffman, P. Johnson, J.-W. Lee, W. Lee, A. Ma, A. Saraf, M. Seneski, N. Shnidman, V. Strumpen, M. Frank, S. Amarasinghe, and A. Agarwal. The raw microprocessor: A computational fabric for software circuits and general-purpose programs. *IEEE Micro*, 22(2):25–35, 2002.

[186] Impulse Accelerated Technologies. Impulsec official web site. http://www.impulsec.com/, 2006.

[187] R. Tessier and W. Burleson. Reconfigurable computing for digital signal processing: A survey. *Journal of VLSI Signal Processing*, 28(1–2):7–27, 2001.

[188] K. Tiensyrjä, Y. Qu, Y. Zhang, M. Cupak, L. Rynders, G. Vanmeerbeeck, K. Masselos, K. Potamianos, and M. Pettissalo. SystemC and OCAPI-xl based system-level design for reconfigurable systems-on-chip. In *Proceedings of the International Forum on Specification & Design Languages (FDL)*, pages 428–429, September 2004.

[189] N. Tredennick. The case for reconfigurable computing. *Microprocessor Report*, 10(10):25–27, 1996.

[190] C.-H. Tseng and P.-A. Hsiung. A UML-based design flow and partitioning methodology for dynamically reconfigurable systems. In *Proceed-*

ings of the IFIP International Conference on Embedded and Ubiquitous Computing (EUC2005), Lecture Notes in Computer Science (LNCS), volume 3824, pages 479–488, December 2005.

[191] C. W. Tseng. *XAPP452 Spartan-3 Advanced Configuration Architecture.* Xilinx Inc., 1.0 edition, December 2004.

[192] M. Ullmann, M. Hübner, B. Grimm, and J. Becker. An FPGA runtime system for dynamical on-demand reconfiguration. In *Proceedings of the 18th International Parallel and Distributed Processing Symposium*, 2004.

[193] F. Vahid and T. Givargis. *Embedded Systems Design – A Unified Hardware/Software Introduction.* John Wiley & Sons Inc., 2002.

[194] M. Vasilko. DYNASTY: A temporal floorplanning based CAD framework for dynamically reconfigurable logic systems. In *Proceedings of the 9th International Workshop on Field-Programmable Logic and Applications*, pages 124–133, 1999.

[195] M. Vasilko and D. Ait-Boudaoud. Architectural synthesis techniques for dynamically reconfigurable logic. In *Proceedings of the 6th International Workshop on Field-Programmable Logic*, pages 290–296, 1996.

[196] M. Vasilko and G. Benyon-Tinker. Automatic temporal floorplanning with guaranteed solution feasibility. In *Proceedings of the The Roadmap to Reconfigurable Computing, 10th International Workshop on Field-Programmable Logic and Applications*, pages 656–664, 2000.

[197] N. S. Voros and K. Masselos. *System Level Design of Reconfigurable Systems-on-Chip.* Springer, 2005.

[198] B. Wade. *Application Notes 416. Using an RPM Grid Macro to Control Block RAM-to-FF Timing.* Xilinx Inc., August 2002.

[199] H. Walder and M. Platzner. Online scheduling for block-partitioned reconfigurable devices. In *Proceedings of the Conference on Design, Automation and Test in Europe (DATE03)*, volume 1, pages 10290–10295, March 2003.

[200] H. Walder and M. Platzner. Reconfigurable hardware operating systems: From concepts to realizations. In *Proceedings of the 3rd International Conference on Engineering of Reconfigurable Systems and Applications (ERSA)*, pages 284–287, June 2003.

[201] H. Walder, C. Steiger, and M. Platzner. Fast online task placement on FPGAs: Free space partitioning and 2D-hashing. In *Proceedings on Parallel and Distributed Processing Symposium*, volume 17, page 8. IEEE CS Press, April 2003.

[202] A. Weisensee and D. Nathan. A self-reconfigurable computing platform hardware architecture, 2004.

[203] M. Wermelinger, A. Lopes, and J. Fiadeiro. A graph based architectural (re)configuration language. In *Proceedings of the Joint European Software Engineering Conference and Symposium on the Foundations of Software Engineering (ESEC/FSE)*. ACM Press, 2001.

[204] G. Wigley and D. Kearney. The development of an operating system for reconfigurable computing. In *Proceedings of the 9th IEEE Symposium on Field-Programmable Custom Computing Machines*, pages 249–250, 2001.

[205] G. Wigley and D. Kearney. The first real operating system for reconfigurable computers. In *Proceedings of the 6th Australasian Computer Systems Architecture Conference*, pages 130–137, 2001.

[206] G. Wigley and D. Kearney. The management of applications for reconfigurable computing using an operating system. In *Proceedings of the 7th Asia-Pacific Computer Systems Architectures Conference*, pages 73–81, January 2002.

[207] G. Wigley, D. Kearney, and M. Jasiunas. ReConfigME: a detailed implementation of an operating system for reconfigurable computing. In *Proceedings of the 20th International Symposium on Parallel and Distributed Processing*, page 8, April 2006.

[208] G. Wigley, D. Kearney, and D. Warren. Introducing ReConfigME: An operating system for reconfigurable computing. In *Proceedings of the 12th IEEE International Conference on Field Programmable Logic and Applications (FPL 2002)*, pages 687–697, 2002.

[209] J. A. Williams and N. W. Bergmann. Embedded Linux as a platform for dynamically self-reconfiguring systems-on-chip. In *Proceedings of International Conference on Engineering of Reconfigurable Systems and Algorithms*, pages 163–169. CSREA Press, June 2004.

[210] W. L. Winston. *Introduction to Mathematical Programming: Applications and Algorithms.* Duxbury Resource Center, 2003.

[211] Xilinx. *Partial Reconfiguration Early Access Software Tool.* Xilinx Inc. http://www.xilinx.com/support/prealounge/protected/index.htm.

[212] Xilinx. ML401/ML402/ML403 Evaluation Platform User Guide, UG080 (v2.5), 2006

[213] Xilinx. UG208 – Early Access Partial Reconfiguration User Guide, 2006.

[214] Xilinx. ML310 User Guide, 2007.

[215] Inc. Xilinx. Microblaze processor reference guide. *Embedded Development Kit, EDK 8.2i. Xilinx User Guide v6.0*, June 2006.

[216] Xilinx Inc. *Virtex-II Pro Data Sheet Virtex-II Pro^{TM} Platform FPGA Data Sheet*, 2003.

[217] Xilinx Inc. *Spartan-3 FPGA Family: Complete Data Sheet.* Xilinx Inc., 2007.

[218] Xilinx Inc. *UG012 Virtex-II Pro FPGA User Guide.* Xilinx Inc., 4.2 edition, November 2007.

[219] Xilinx Inc. *Virtex-4 Family Overview.* Xilinx Inc., 2007.

[220] X. Yan and J. Han. Closegraph: mining closed frequent graph patterns. In *KDD '03: Proceedings of the Ninth ACM SIGKDD International Conference on Knowledge Discovery and Data Mining*, pages 286–295. ACM Press, 2003.

[221] P. Yang, C. Wong, P. Marchal, F. Catthoor, D. Desmet, D. Verkest, and R. Lauwereins. Energy-aware runtime scheduling for embedded multiprocessor SoCs. *IEEE Journal on Design and Test of Computers*, 18(5):46–58, September 2001.

[222] M. J. Zaki. Efficiently mining frequent trees in a forest: algorithms and applications. *IEEE Transactions on Knowledge and Data Engineering*, 17(8):1021–1035, 2005.

Index

N

O

Q

R